United States Nuclear Regulatory Commission

Protecting People and the Environment

NUREG/CR-7158
ORNL/TM-2012/261

I0493880

Review and Prioritization of Technical Issues Related to Burnup Credit for BWR Fuel

Office of Nuclear Regulatory Research

AVAILABILITY OF REFERENCE MATERIALS
IN NRC PUBLICATIONS

NRC Reference Material

As of November 1999, you may electronically access NUREG-series publications and other NRC records at NRC's Public Electronic Reading Room at http://www.nrc.gov/reading-rm.html. Publicly released records include, to name a few, NUREG-series publications; *Federal Register* notices; applicant, licensee, and vendor documents and correspondence; NRC correspondence and internal memoranda; bulletins and information notices; inspection and investigative reports; licensee event reports; and Commission papers and their attachments.

NRC publications in the NUREG series, NRC regulations, and Title 10, "Energy," in the *Code of Federal Regulations* may also be purchased from one of these two sources.
1. The Superintendent of Documents
   U.S. Government Printing Office Mail Stop SSOP
   Washington, DC 20402–0001
   Internet: bookstore.gpo.gov
   Telephone: 202-512-1800
   Fax: 202-512-2250
2. The National Technical Information Service
   Springfield, VA 22161–0002
   www.ntis.gov
   1–800–553–6847 or, locally, 703–605–6000

A single copy of each NRC draft report for comment is available free, to the extent of supply, upon written request as follows:
Address: U.S. Nuclear Regulatory Commission
         Office of Administration
         Publications Branch
         Washington, DC 20555-0001
E-mail: DISTRIBUTION.RESOURCE@NRC.GOV
Facsimile: 301–415–2289

Some publications in the NUREG series that are posted at NRC's Web site address http://www.nrc.gov/reading-rm/doc-collections/nuregs are updated periodically and may differ from the last printed version. Although references to material found on a Web site bear the date the material was accessed, the material available on the date cited may subsequently be removed from the site.

Non-NRC Reference Material

Documents available from public and special technical libraries include all open literature items, such as books, journal articles, transactions, *Federal Register* notices, Federal and State legislation, and congressional reports. Such documents as theses, dissertations, foreign reports and translations, and non-NRC conference proceedings may be purchased from their sponsoring organization.

Copies of industry codes and standards used in a substantive manner in the NRC regulatory process are maintained at—
       The NRC Technical Library
       Two White Flint North
       11545 Rockville Pike
       Rockville, MD 20852–2738

These standards are available in the library for reference use by the public. Codes and standards are usually copyrighted and may be purchased from the originating organization or, if they are American National Standards, from—
       American National Standards Institute
       11 West 42nd Street
       New York, NY 10036–8002
       www.ansi.org
       212–642–4900

Legally binding regulatory requirements are stated only in laws; NRC regulations; licenses, including technical specifications; or orders, not in NUREG-series publications. The views expressed in contractor-prepared publications in this series are not necessarily those of the NRC.

The NUREG series comprises (1) technical and administrative reports and books prepared by the staff (NUREG–XXXX) or agency contractors (NUREG/CR–XXXX), (2) proceedings of conferences (NUREG/CP–XXXX), (3) reports resulting from international agreements (NUREG/IA–XXXX), (4) brochures (NUREG/BR–XXXX), and (5) compilations of legal decisions and orders of the Commission and Atomic and Safety Licensing Boards and of Directors' decisions under Section 2.206 of NRC's regulations (NUREG–0750).

United States Nuclear Regulatory Commission

Protecting People and the Environment

NUREG/CR-7158
ORNL/TM-2012/261

Review and Prioritization of Technical Issues Related to Burnup Credit for BWR Fuel

Manuscript Completed: November 2012
Date Published: February 2013

Prepared by
D. E. Mueller
S. M. Bowman
W. J. Marshall
J. M. Scaglione

Oak Ridge National Laboratory
Managed by UT-Battelle, LLC
Oak Ridge, TN 37831-6170

M. Aissa, NRC Project Manager

NRC Job Code V6061

Prepared for
Division of Systems Analysis
Office of Nuclear Regulatory Research
U.S. Nuclear Regulatory Commission
Washington, DC 20555-0001

ABSTRACT

This report has been prepared to support technical discussion of and planning for future research supporting implementation of burnup credit for boiling-water reactor (BWR) spent fuel storage in spent fuel pools and storage and transport cask applications. The review and discussion in this report are based on knowledge and experience gained from work performed in the United States and other countries, including experience with burnup credit for pressurized-water reactor (PWR) spent fuel. Relevant physics and analysis phenomena are identified, and an assessment of their importance to burnup credit implementation is given. Results from sensitivity studies of some of the key phenomena are presented.

The work presented in this report is primarily a sensitivity study designed to identify and rank phenomena and parameters important to BWR burnup credit methodology. This work is an extension of the work reported in NUREG/CR-7157, *Computational Benchmark for Estimated Reactivity Margin from Fission Products and Minor Actinides in BWR Burnup Credit*, which defines the baseline BWR spent fuel cask model used in this report and provides estimates for the reactivity margin associated with fission products and minor actinides. All calculations supporting this work were performed using the SCALE 6.1 code system with the 238-neutron energy group ENDF/B-VII-based nuclear data library.

Discussion of and recommendations for future work supporting implementation of BWR burnup credit beyond the currently used peak reactivity method are provided. A high priority is recommended for development of guidance for identification and use of axial burnup distribution data, for treatment of axial moderator density distributions, and for treatment of control blade usage during depletion calculations. In addition to these higher priority items, several medium and lower priority items are identified and discussed.

CONTENTS

LIST OF FIGURES

LIST OF TABLES

ACKNOWLEDGMENTS

This work was performed under contract with the Office of Nuclear Regulatory Research, U.S. Nuclear Regulatory Commission (NRC). The authors thank M. Aissa, the NRC Project Manager, D. R. Algama and R. Y. Lee of the Office of Nuclear Research (RES), T. T. Nakanishi and K. A. L. Wood of the Office of Nuclear Reactor Regulation (NRR), A. B. Barto and Z. Li of the Office of Nuclear Material Safety and Safeguards (NMSS) for their support and guidance. The careful reviews of the draft manuscript by Brian Ade and Andrew Godfrey are very much appreciated. Finally, the authors are thankful to A. C. Alford for her preparation of the final report.

ACRONYMS

AFP:	Actinides plus fission products; as in actinide plus 16 fission product burnup credit
ANS:	American Nuclear Society
ANSI:	American National Standards Institute
AO:	Actinide-only; as in actinide-only burnup credit
BUC:	burnup credit
BWR:	boiling-water reactor
CRC:	commercial reactor critical
DOE:	U.S. Department of Energy
FP:	fission product
GBC:	generic burnup credit cask; as in GBC-68, a generic burnup credit cask model with a 68 BWR fuel assembly capacity
Gd:	gadolinium
GWd/MTU:	unit of nuclear fuel burnup; gigawatt-days per initial metric ton of uranium
ISG:	interim staff guidance
LWR:	light-water reactor
NRC:	U.S. Nuclear Regulatory Commission
OCRWM:	DOE Office of Civilian Radioactive Waste Management
ORNL:	Oak Ridge National Laboratory
PWR:	pressurized-water reactor
SCCG:	standard cold core geometry
SFP:	spent fuel pool
SNF:	spent nuclear fuel
YMP:	Yucca Mountain Project

1. INTRODUCTION

The concept of taking credit for the reduction in reactivity due to the net consumption of fissile nuclides and creation of neutron absorbing actinides and fission products during reactor operation is commonly referred to as "burnup credit" (BUC). In the past decade, numerous studies have investigated the issues and phenomena associated with the application of burnup credit to pressurized-water reactor (PWR) fuel for storage in spent fuel pools (SFPs) and in storage and transport casks. In contrast, far fewer studies have been performed for boiling-water reactor (BWR) fuel. This disparity has been due to the lack of a perceived need for BWR burnup credit and the complexity in modeling BWR spent nuclear fuel (SNF). SNF storage capacity limitations, degradation of neutron absorber material in spent fuel pools, the lack of progress in establishing a SNF permanent disposal facility, and nuclear power plant license extensions have caused the industry to consider BUC applications for BWR SNF.

For storage of discharged BWR assemblies in spent fuel pools, several approaches have been applied in the United States for criticality safety analyses. The most frequently used analysis methods include defining assembly storage acceptability in terms of (1) maximum lattice average initial ^{235}U enrichment, neglecting credit for integral neutron poisons and for fuel assembly burnup, (2) maximum assembly lattice k_∞ value in an artificial reference configuration (i.e., the standard cold core geometry, SCCG), and (3) defining assembly acceptability on a case-by-case basis in terms of maximum enrichment and minimum gadolinium loading, again neglecting credit for burnup. Although these approaches have in the past apparently provided sufficient operational flexibility and negative reactivity credit for spent fuel pools, the negative reactivity credit associated with fuel burnup is either ignored or essentially limited to the reactivity worth of the burnable absorber. The analyses are not completely independent of lattice design variations such as gadolinium loading.

The most widely used approach for BWR spent fuel pools is based on depletion calculations that are performed to identify the burnup point at which the reactivity in cold conditions is maximized, i.e. the burnup point where the combination of remaining fissile nuclides and integral burnable poison content yields the highest reactivity, for the most reactive lattice to be stored. This burnup point is sometimes referred to as the burnup of the "gadolinium peak" or the "reactivity peak" and, for the criticality analysis limiting cases, typically occurs between 10 and 20 gigawatt-days per initial metric ton of uranium (GWd/MTU) for gadolinium (Gd) poisoned fuel pins where gadolinia (Gd_2O_3) concentrations are 3 to 8% of the fuel pellet mass. This approach ensures an absolute reactivity maximum is used in the supporting criticality analysis; this is analogous to what constitutes the "fresh fuel approach" for PWR fuel (without integral burnable absorbers) in the sense that after demonstrating that such maximum reactivity is acceptable for the storage conditions, no criteria are needed relative to the actual burnup of the fuel assembly. This is a desirable feature of the methodology, as it requires no limits or controls associated with burnup verification. On the other hand, this approach neglects the additional reactivity decrease obtained for higher burnup values, i.e., burnup values higher than that corresponding to the reactivity peak.

Another potentially viable approach is to define assembly acceptability in terms of maximum initial lattice average ^{235}U enrichment and minimum burnup, consistent with the approach generally taken for PWR burnup credit. It may also be possible to use the combination of the maximum assembly lattice k_∞ approach at lower burnup values and the maximum initial enrichment and minimum burnup at burnup values above the peak reactivity burnup. In spent fuel pools (SFPs), expanded burnup credit might be implemented using separate storage

1

regions for low burnup fuel, based on the peak reactivity method, and higher burnup fuel, based on no-gadolinia burnup credit curves.

The NRC issued Interim Staff Guidance 8 (ISG-8), "Burnup Credit in the Criticality Safety Analyses of PWR Spent Fuel in Transport and Storage Casks" in May 1999 [1], providing the first allowance of burnup credit for PWR fuel in casks. Based on technical work performed at ORNL and elsewhere, ISG-8 has undergone three revisions [2, 3, and 4], which have eliminated or lessened a number of the restrictions. However, ISG-8 is specific to PWR fuel, and no such similar guidance permitting burnup credit for BWR fuel in storage and transport has been developed. The regulatory standard review plans for dry cask storage and transport do not permit credit for BWR fuel burnup or fixed (i.e., integral) burnable absorbers. Relevant excerpts from NUREG-1617 [5] include, from Section 6.5.2;

> *For BWR fuel assemblies, NRC staff does not currently allow any credit for burnup of the fissile material or increase in actinide or fission product poisons during irradiation; therefore, the enrichment should be that of the un-irradiated fuel.*

And from Section 6.5.3.2;

> *. . . because of differences in net reactivity due to depletion of fissile material and burnable poisons, no credit should be taken for burnable poisons in the fuel.*

1.1   PURPOSE

Numerous studies have been performed nationally and internationally to develop a detailed understanding of the issues and phenomena associated with PWR burnup credit. The understanding resulting from these studies strengthened the bases supporting allowance of burnup credit for PWR fuel. However, a technical basis for BWR burnup credit has not been developed. This report has been created to support identification and establishment of the requisite understanding and lay a foundation for development of a technical basis for allowance of expanded burnup credit for BWR fuel in both spent fuel pools and storage and transport casks.

1.2   BACKGROUND AND PRIOR WORK

Contributing factors to the lack of impetus for research in BWR burnup credit have included the availability of needed storage space in spent fuel pools, expectations of long-term storage and disposal capability, and complexity in the modeling and analysis of BWR spent nuclear fuel (SNF). These modeling complexities include radial and axial variations in fuel enrichment, burnable absorber content, extensive use of control blades during operation, and significant axial moderator density variation due to a combination of two-phase flow and varying core flow.

Some studies have been performed to support burnup credit for BWR fuel, but the issues and phenomenon have not been addressed in a thorough or systematic manner for relevant dry cask storage and transport systems as has been done for PWR fuel. Relevant BWR studies are summarized below.

Reference 6, ORNL/M-6155, was published in August 1999. This work evaluated trends in BWR spent fuel assembly k_∞ values as a function of various burnup/initial enrichment and cooling time combinations for the Yucca Mountain Project (YMP).

Reference 7, ORNL/TM-1999/193, was published in October 2000. This work investigated various calculational modeling issues that are associated with BWR fuel depletion and are relevant to burnup credit.

Reference 8, NUREG/CR-6665 (ORNL/TM-1999/303), was published in February 2000. This report was prepared to review relevant background information and provide technical discussions that were intended to help initiate a PIRT (Phenomena Identification and Ranking Tables) process for use of burnup credit in light-water reactor spent fuel storage and transport cask applications. This report includes some discussion specific to BWR spent fuel burnup credit.

Reference 9 was prepared by K.W. Cummings and S. E. Turner for the *2001 ANS Embedded Topical Meeting on Practical Implementation of Nuclear Criticality Safety* in Reno, NV in November of 2001. This paper describes an implementation of BWR fuel storage criticality analysis utilizing the standard cold core geometry k_∞ as the parameter defining storage acceptability.

Reference 10 was prepared by C. Casado, J. Sabater, and J. F. Serrano for the *IAEA International Workshop on Advances in Applications of Burnup Credit for Spent Fuel Storage* in Cordoba, Spain in October 2009. This paper proposes a simple conservative method for the definition of the peak reactivity point and isotopic inventory calculation for BWR criticality applications.

Reference 11 was prepared by J. Huffer and J. M. Scaglione for the Office of Civilian Radioactive Waste Management (OCRWM) for the YMP in October 2003. The report presented the methods and results from an effort to use BWR spent fuel radiochemical assay data to determine bias and uncertainty associated with BWR spent fuel composition calculations.

Reference 12 was prepared by J. M. Scaglione for the OCRWM for the YMP in August 2004. The report presented the method used to determine the required minimum burnup as a function of initial BWR assembly enrichment that would permit loading of spent nuclear fuel into a BWR 44-assembly waste package.

Reference 13 was prepared by J. Huffer for the OCRWM for the YMP in September 2004. The report develops and evaluates a method for determining conservative axial burnup, power, water density and temperature profiles for BWR spent fuel assembly modeling for burnup credit.

Burnup credit is typically considered for BWR fuel in SFP criticality analyses, albeit in a limited and different approach compared to that used for PWR fuel. Although the approaches described in Section 1.0 have been used in BWR SFP analyses for demonstrating sufficient negative reactivity credit, the negative reactivity credit is essentially limited to the reactivity worth of fuel depletion to the maximum reactivity point and the residual burnable absorber still present at that point, and the analyses are not completely independent of lattice design variations such as gadolinium loading. Credit has not been taken for further reactivity reduction due to fuel depletion beyond the peak reactivity point. Hence, there is a need for a simplified modeling approach that defines assembly acceptability in terms of maximum initial enrichment and minimum burnup, consistent with the approach generally taken for PWR burnup credit.

Modern BWR fuel assemblies make heavy use of burnable absorbers, have heterogeneous time-dependent moderator densities, and may be adjacent to inserted control blades during normal operation. In many cases, details of the operating history of benchmark spent fuel assay samples are considered commercial proprietary information and are not well documented in public sources. The lack of adequate documentation for modern BWR assembly designs and operational void distribution, control blade usage, and fuel temperature histories for the spent fuel assemblies has been a major impediment to the availability of quality benchmark data for fuel depletion code validation.

During the process of preparing the Yucca Mountain license application, a considerable amount of information pertaining to BWR fuel and burnup credit was made available [14], and recent focus domestically and internationally on BWR burnup credit has resulted in a fair amount of detailed data becoming available. In addition to detailed assembly design information, the relevant ranges of depletion parameters that can affect neutron spectrum need to be known.

Discussion and descriptions of relevant design and operating data and their potential impact on BWR burnup credit are provided in the following sections. The current effort documented in this report is directed toward establishing the requisite understanding and technical basis for allowance of burnup credit for BWR fuel in storage and transportation casks, and leveraging current accepted practices used in SFP criticality safety evaluations. The work presented in this report builds on the foundation of work presented in the companion NUREG/CR report *Computational Benchmark for Estimated Reactivity Margin from Fission Products and Minor Actinides in BWR Burnup Credit*, NUREG/CR-7157 [15], which defines the reference or base-line model used in this follow-on report and provides estimates for the reactivity margin associated with minor actinides and fission products that may be credited in BWR burnup credit.

2. BOILING WATER REACTOR CHARACTERISTICS

2.1 REACTOR DESIGN AND OPERATIONS

Burnup credit is credit for reduction in fuel assembly reactivity that results from use of the fuel in the reactor. As fissions in the fuel generate heat during reactor operation, there is a net reduction in fissile material and actinides and fission products are created. The changes in the fuel compositions depend not only on the total number of fissions that occur, but also on variations in the energy spectrum of the neutron flux to which the fuel is exposed. Neutron spectral shifts are associated with variables such as moderator density, insertion or withdrawal of control blades, and depletion of gadolinium poison in some fuel rods. Accurate simulation of the neutron spectrum by correctly modeling these variables is needed to correctly calculate the spent fuel isotopic concentrations during fuel depletion. Neutron energy-spectrum shifts to higher energies lead to increased buildup of actinides, including some fissile nuclides such as ^{239}Pu and ^{241}Pu that increase reactivity and other actinides that are primarily neutron absorbers that reduce reactivity. The shift to higher energies also results in reduced depletion of important fission product nuclides that are primarily thermal neutron absorbers. Neutron energy-spectrum shifts to lower energies lead to a reduced buildup of actinides and increased depletion of thermal neutron absorbing fission products. By common nuclear engineering convention, a shift of a neutron spectrum to higher energies is also referred to as a hardening of the spectrum, while a shift to lower energies is sometimes referred to as a softening of the neutron spectrum.

The use of control blades and integral absorbers affects the fuel composition by absorbing thermal neutrons, thus hardening the neutron spectrum. When control blades are not inserted, the space is filled with water, adding moderator between assemblies, which results in a softer, or more thermal, neutron spectrum. An additional complicating effect associated with control blade insertion is that the power density in the fuel near the control blade is reduced. The reduced power density results in an increase in water density and neutron moderation, somewhat softening the neutron spectrum. To simplify criticality analyses, some analysts have in the past utilized worst case modeling assumptions that include assuming concurrent existence of conditions that cannot coexist. For example, one could defend the conservative approach of simultaneously assuming control blade insertion and a conservatively bounding low moderator density during fuel depletion. A more realistic analysis approach would assume a minimum water density that is consistent with an assembly next to a control blade.

Continuous control of a BWR is typically accomplished by controlling the rate of coolant flow through the core. Core flow rate adjustments affect the neutron energy spectrum and fuel composition by changing the effective density of the water flowing through the core. The effective density is the average density of the coolant, including consideration of both the density of the saturated or sub-cooled water and the steam voids introduced by boiling. If the reactor power level is held constant, decreasing the flow rate results in a larger steam or "void" fraction, thereby reducing the effective water density. Water density reduction results in reduced neutron moderation and a hardening of the neutron spectrum. Conversely, increasing the flow rate sweeps the steam voids out of the reactor more quickly, increasing neutron moderation and softening the neutron spectrum.

Due to the two-phase flow present in BWRs at full reactor power, the effective water density around the fuel pins changes from about 0.7 g/cm^3 to values approaching 0.2 g/cm^3 as the water moves from the bottom to the top of the core. For comparison purposes, in a PWR the water density typically varies from around 0.75 g/cm^3 at the bottom of the core to 0.65 g/cm^3 at the top of the core. Consequently, consideration of the axial dependence of the fuel depletion environment is significantly more important for BWRs than for PWRs. Discussion of the impact of the depletion parameters used on the criticality safety analysis is provided in Section 3.

2.2   FUEL ASSEMBLY DESIGN AND OPERATING DATA

BWR fuel assembly lattices have considerable variability. General lattice designs exist for a variety of array configurations including 6 × 6, 7 × 7, 8 × 8, 9 × 9, and 10 × 10, as well as a quad-channel design with an internal water cross. Within each of these different lattice configurations exist other design features, including differences in numbers, size and placement of water rods, numbers and gadolinia content of Gd fuel rods (i.e., fuel rods with some fuel pellets containing a mixture of UO_2 and Gd_2O_3), Gd fuel rod axial and radial loading patterns, variation in radial and axial ^{235}U enrichments, use of part-length rods, which come in different lengths, fuel channel design, and the use of axial blankets. Each of these design features in conjunction with different reactor operating strategies will impact fuel assembly residual reactivity at discharge. In addition, multiple BWR plants have a variety of lattice designs currently stored in the spent fuel pools, which provides an additional complexity to criticality safety evaluations for spent fuel storage and transportation. An 8 x 8 lattice containing one large central water hole is illustrated in Figure 1 to show some of the design features. This graphic was extracted from DOE/RW-0573 [14].

Relevant design and operating data for several of the lattice configurations can be drawn from publicly available reports that contain detailed information for performing commercial reactor criticality (CRC) analyses. This information can be used in evaluations with realistic BWR assembly and lattice designs to determine the influence of features such as integral burnable absorbers, reduced enrichment and/or natural uranium blankets, part-length fuel rods, and control blade insertion histories. Three BWR CRC reports are available as follows:

(1) *Summary Report of Commercial Reactor Criticality Data for Grand Gulf Unit 1* [16]. This report provides fuel assembly design information for Cycles 2 through 8 of the Grand Gulf Unit 1 reactor. Material and geometry data for the fuel assembly components are included. The fuel assembly ^{235}U weight percentage enrichments and gadolinia (Gd_2O_3) enrichments for each fuel design of Cycles 2 through 8 are also presented. Fuel assembly designs include: a Siemens Power Corporation (SPC) 8 × 8 lattice configuration with two water rods, one of which acts as a spacer capture rod; and a SPC 9×9-5 which is a 9 × 9 array with 5 water rods, axially zoned enrichments, and integral burnable absorbers.

(2) *Summary Report of Commercial Reactor Criticality Data for LaSalle Unit 1* [17]. This report provides fuel assembly design information for Cycles 4 through 8 of the LaSalle Unit 1 reactor. Material and geometry data for the fuel assembly components are included. The fuel assembly ^{235}U weight percentage enrichments and gadolinia (Gd_2O_3) enrichments for each fuel design of Cycles 4 through 8 are also presented. The fuel assembly design was a General Electric (GE) 8 × 8NB design consisting of an 8 × 8 array with a central, single large-diameter water rod that occupies four fuel rod positions, axially zoned enrichment, and integral burnable absorbers. In addition, this assembly design has a 12-inch natural uranium blanket at the top.

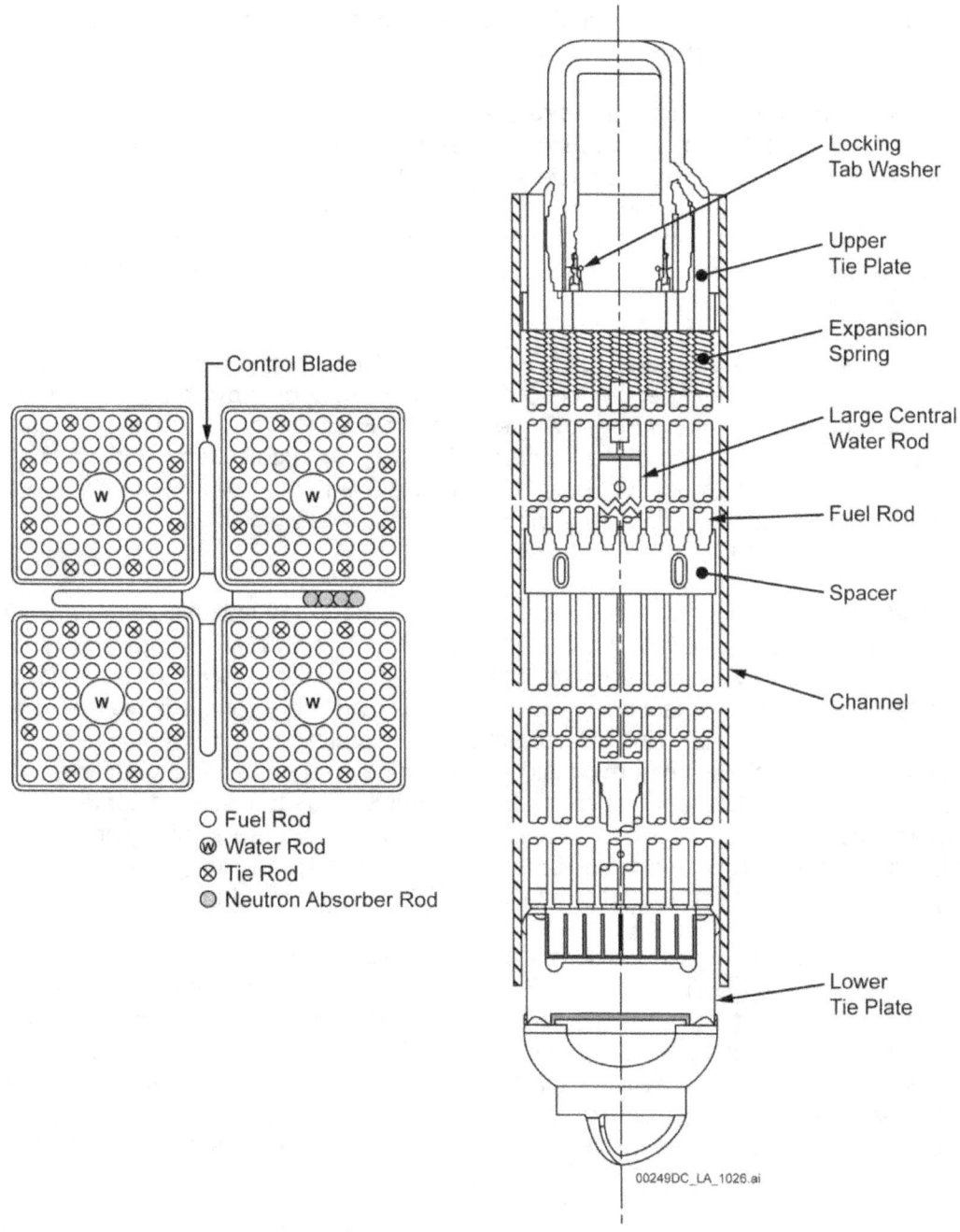

Figure 1. Example 8 x 8 fuel assembly configuration [14].

(3) *Summary Report of Commercial Reactor Criticality Data for Quad Cities Unit 2* [18]. This report provides fuel assembly design information for Cycles 9 through 14 of the Quad Cities Unit 2 reactor. Material and geometry data for the fuel assembly components are included. The fuel assembly ^{235}U weight percentage enrichments and gadolinia (Gd_2O_3) enrichments for each fuel design of Cycles 9 through 14 are also presented. The fuel assembly designs consisted of: the GE8×8EB design, which contains two water rods; the GE8 fuel design which contains four water rods, axially zoned enrichments, and gadolinia rods; the GE8×8NB design; and a GE8×8NB-3 design, which is similar to the

GE8×8NB design but contains fuel channel thickness variations (thinner faces than corners) and flow directors.

The BWR CRC data provided in the three references above should be used with care. The information describing the reactor operating environment, such as local moderator density, fuel temperatures, and power density, are best-estimate data that are inferred from a combination of measurements and nodal simulator code (e.g. SIMULATE) calculations. Little information is provided concerning the sources of the data and the data uncertainties. However, these data are still very useful for exploring the impact of reactor operating conditions on BWR burnup credit.

One of the more modern BWR fuel assembly designs, a GE14 assembly, consisting of a 10 × 10 lattice with axially vanished regions and two water rods occupying eight fuel rod lattice locations is described in *Optimum Boiling Water Reactor Fuel Design Strategies to Enhance Reactor Shutdown by the Standby Liquid Control System* [19]. Axially "vanished" regions result from use of part length rods present at only lower axial elevations. A detailed description showing the axial and radial distribution is presented in Figure 2, which was generated using data from and is similar to Fig. 2-2 of Reference 19.

Another source available for information regarding modern 10 × 10 lattice configurations is *Peak Reactivity Characterization and Isotopic Inventory Calculations for BWR Criticality Applications* [10]. This source summarizes a burnup credit study performed for a 10 × 10 lattice with three different axial configurations investigating development of a simple conservative method for defining the peak reactivity point and isotopic inventory calculation for BWR criticality applications. The relevant parameters considered included core conditions, gadolinium rod locations, and the initial fissile enrichment distribution, as well as the isotopic content for the spent fuel and the axial burnup shape of the fuel assembly. The three axial regions consisted of a full lattice specification, a vanishing region zone, and a second vanishing region zone, indicating that the assembly consisted of multiple lengths of part-length fuel rods.

The burnup credit criticality analysis needs to show that all permitted lattices and other fuel assembly design variations will meet criticality safety k_{eff} limits. This has typically been accomplished for PWR burnup credit by identifying the most reactive lattice, including the most reactive set of design variations, or a clearly bounding fictitious lattice. The analyst and the reviewer need to be aware that the most reactive lattice may vary with burnup. One lattice may be bounding at lower burnups, while a different lattice may become relatively more reactive at higher burnups. The impact of fuel assembly design variation is discussed further in Section 3.

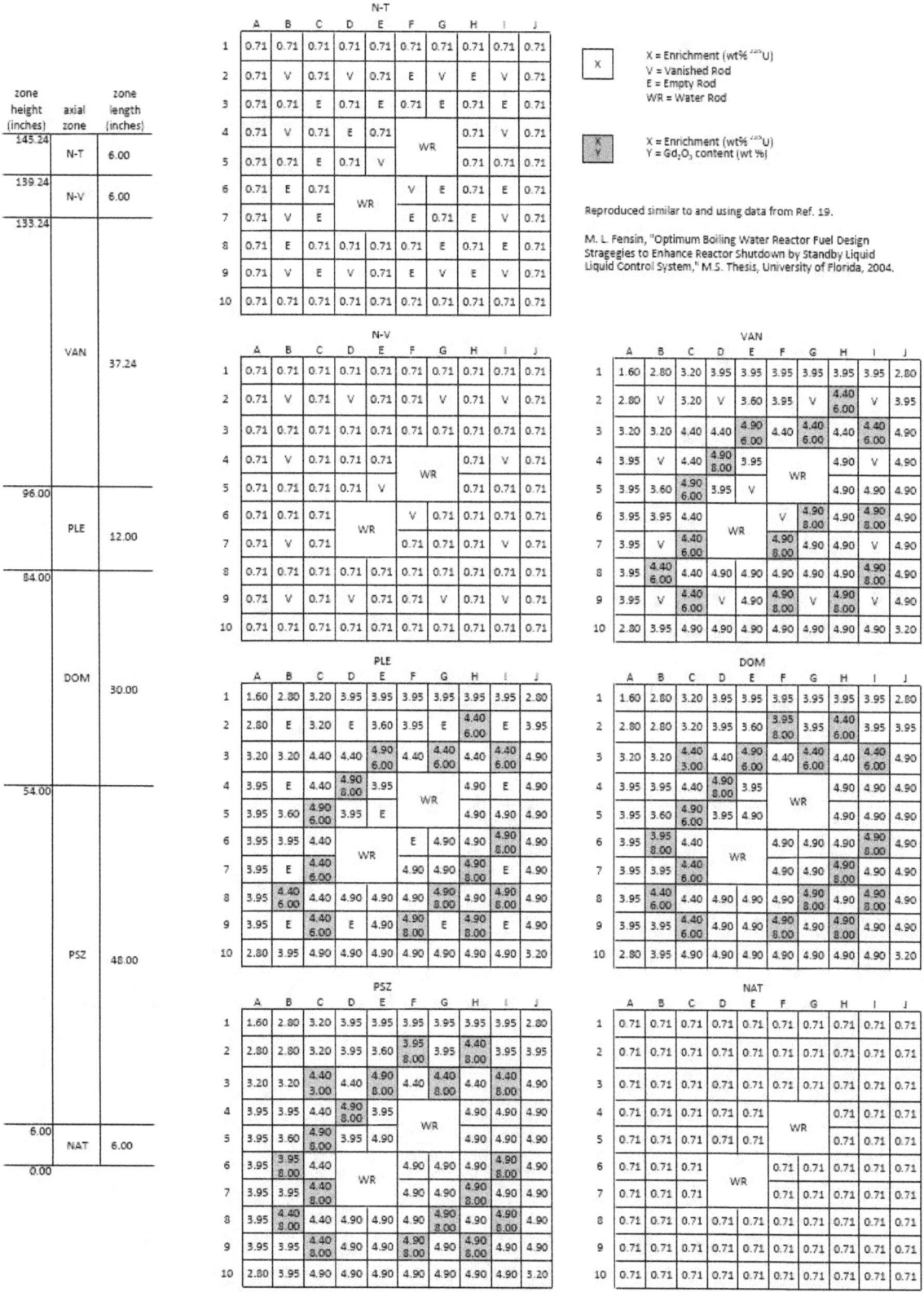

Figure 2.   Typical GE14 10x10 lattice [19].

9

3. BOILING WATER REACTOR BURNUP CREDIT

3.1 BURNUP CREDIT ANALYSIS APPROACH

Currently, there is no standard or generally accepted approach for nuclear criticality safety analyses for BWR fuel crediting burnup beyond peak reactivity. It seems reasonable to assume that the currently-used peak reactivity analysis technique was adopted because the number of assembly design variations was large enough to render performance of criticality analysis for each variation impractical and not crediting gadolinium in the low-burnup fuel yielded unacceptably high k_{eff} values.

A simple expansion of the peak reactivity analysis technique to include credit for burnup beyond the peak reactivity point is not appropriate because the technique uses a two-dimensional model. NUREG/CR-6801 [20] provides discussion and recommendations concerning the impacts of the axial burnup distribution on PWR burnup credit. As shown in Figure 3, which is reproduced from Figure 2 of NUREG/CR-6801 for PWR burnup credit, it is non-conservative to ignore the axial burnup distribution at burnup values above 10 GWd/MTU. As will be discussed later in this section, there are several reasons why it will be important to model the axial burnup distribution for BWR burnup credit. Because the peak reactivity analysis technique utilizes a two-dimensional lattice calculation to quantify the standard cold core geometry (SCCG) k_{∞}, this technique cannot include the influence of axial burnup distribution on lattice reactivity at higher burnup values.

Alternative BWR burnup credit analysis approaches are needed to support burnup credit beyond the peak reactivity burnup point. It is proposed that BWR burnup credit analysis could be performed in a manner similar to PWR burnup credit by omitting consideration of the gadolinia in the fuel pellets. This would necessitate the use of bounding axial burnup profiles and include consideration of the axially and radially varying design features and depletion parameters. It may be possible to simplify modeling of the axially and radially varying design features by modeling all fuel pins using the highest lattice average initial ^{235}U enrichment and by using the most reactive axial lattice for the entire length of the assembly or for the entire length between the axial blankets used by modern BWRs. Additional study would be needed to support use of this proposed BWR burnup credit technique.

At low burnups, this proposed technique would have limited usefulness in fuel storage rack analyses because ignoring the gadolinia would result in unacceptably high fuel storage rack k_{eff} values. To address this issue, fuel storage rack analyses might utilize the peak reactivity analysis method at low burnup values or in a separate fuel storage rack region and the no-gadolinia burnup credit analysis technique at burnups above the peak reactivity burnup point or in a separate fuel storage rack region. Additional study would be needed to support use of a combined peak reactivity plus no-Gd hybrid BWR burnup credit technique.

It may also be possible to implement full range burnup credit that includes credit for gadolinia in the Gd-bearing fuel rods and realistic consideration of axially-dependent assembly design features. The primary advantages to this approach would be increased Gd credit at burnups below the peak reactivity burnup and a consistent burnup credit approach over the entire fuel assembly burnup range. The primary disadvantage is the need to address the complexities associated with fuel assembly designs that include consideration of the full range of assembly design features.

This report does not contain a recommendation for a single BWR burnup credit analysis approach. The selection of a BWR burnup credit analysis approach by an applicant will depend on the amount of burnup credit needed, the complexity of the fuel and storage system designs, and the ability of operations staff to implement complex limits and controls. Future BWR burnup credit research and development efforts should be designed to be useful for a range of analysis methods.

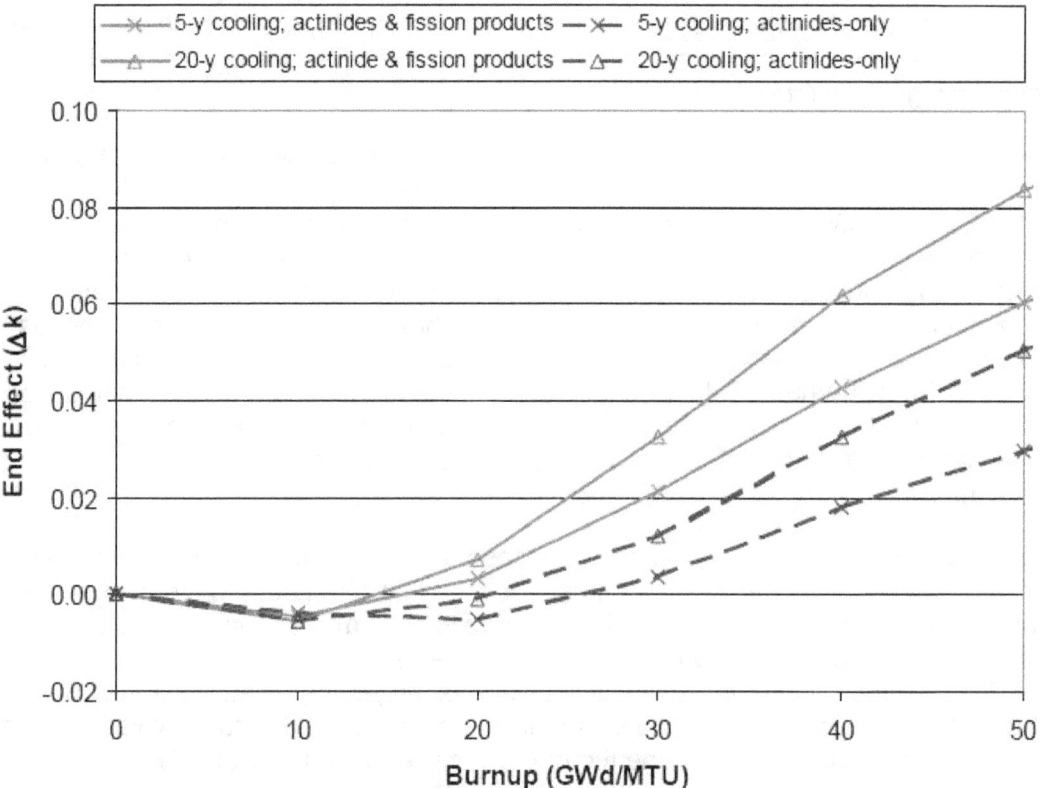

Figure 3. Example of the end effect as a function of burnup for two cooling times with and without fission products present.

3.2   BWR BURNUP CREDIT CALCULATIONS

Numerous burned fuel composition and k_{eff} calculations were performed to support the discussion and analysis presented in this report. These sensitivity calculations were performed using variations of the BWR generic burnup credit cask model, GBC-68, described in NUREG/CR-7157. The computer codes and nuclear data used are described in Section 3.2.1 and the fuel depletion, fuel assembly, and cask models are described in Section 3.2.2.

3.2.1 Computer Codes and Nuclear Data Used

Some of the sections following this section include calculated sensitivity information. All calculations were performed using the publicly released version of SCALE 6.1 [21] and the ENDF/B-VII 238 neutron energy group library distributed with SCALE 6.1. This section provides brief descriptions of the computer codes and nuclear data used. Section 3.2.2 provides additional detail concerning how these computer codes and data were used.

TRITON

TRITON is a multipurpose SCALE control module that can be used for neutron transport, fuel depletion calculations, and sensitivity/uncertainty analysis for reactor physics applications. A complete description of the TRITON control module is provided in Section T01 of the SCALE 6.1 manual [21]. The work presented in this report utilized the TRITON *T-DEPL* sequence to generate ORIGEN-ARP libraries for use by the SCALE STARBUCS control module, which is described below, and to generate burned fuel compositions for direct use in SCALE CSAS5/KENO-V.a k_{eff} calculations. The *T-DEPL* sequence uses NEWT, a multigroup discrete-ordinates neutron two-dimensional transport code, and the ORIGEN-S depletion module to simulate fuel depletion.

Figure 4 is a simplified flow chart showing how the TRITON *T-DEPL* sequence was used for the BWR BUC sensitivity studies. The user input file describes the two-dimensional fuel assembly geometry and depletion conditions and defines materials for specific geometric regions. The input file also specifies the number of days the fuel is depleted and cooled following depletion. Incorporation of problem-dependent resonance self-shielding contributions to the cross sections is accomplished by BONAMI, CENTRM and PMC. NEWT is then executed using the problem-dependent cross sections to calculate the spatially-dependent flux distribution. TRITON passes the neutron fluxes and fuel compositions to COUPLE and ORIGEN-S, which calculate updated burned fuel compositions that are returned to TRITON. ORIGEN-S also produces ORIGEN-ARP libraries that may be used in ORIGEN-ARP calculations employed by STARBUCS to rapidly generate burned fuel compositions. At the end of the depletion step, TRITON generates a set of fuel composition input data in the SCALE Standard Composition block format. The burned fuel composition data may be used directly in CSAS5 calculations of the GBC-68 cask loaded with burned fuel. If additional burnup steps are needed, TRITON cycles back to update the resonance-corrected problem-dependent cross sections and continues looping through cross section preparation, flux distribution calculation, and burned fuel composition calculations until the requested depletion is completed.

Multiple *T-DEPL* calculations with varying initial enrichments and moderator densities were performed to generate problem-dependent ORIGEN-ARP libraries to be used by STARBUCS to rapidly calculate burned fuel compositions. For analyses requiring a very large number of fuel compositions, such as the analysis presented in this report, the ability to rapidly generate burned fuel compositions is invaluable.

Note that each ORIGEN-ARP library is problem-dependent. This means that the library should be used only for similar lattice designs with initial enrichment, final burnup, and moderator density that are consistent with the library and depleted with similar reactor conditions. For example, it would be inappropriate to use a library to model fuel that was depleted at a significantly different power density or fuel temperature. The effects of these depletion conditions are included in the problem-specific library. Consequently, sensitivity studies of some parameters, such as fuel temperature and power density, require generation of either additional ORIGEN-ARP libraries at the varied conditions or direct calculation of problem-specific fuel compositions using TRITON with the varied reactor depletion conditions.

In the work presented in this report, some of the sensitivity studies were performed using STARBUCS. Other calculations were performed by inserting burned fuel compositions calculated by TRITON for multiple fuel assembly conditions directly into CSAS5 k_{eff} calculations. For example, the sensitivity of k_{eff} to variations in reactor moderator density, initial enrichment, and post-irradiation cooling time were calculated using STARBUCS, because these parameters can be varied in ORIGEN-ARP. The sensitivities of k_{eff} to fuel temperature and reactor power density were calculated using TRITON calculated fuel compositions in CSAS5 k_{eff}

calculations, because these parameters cannot be varied in ORIGEN-ARP (they are fixed in the generation of the libraries).

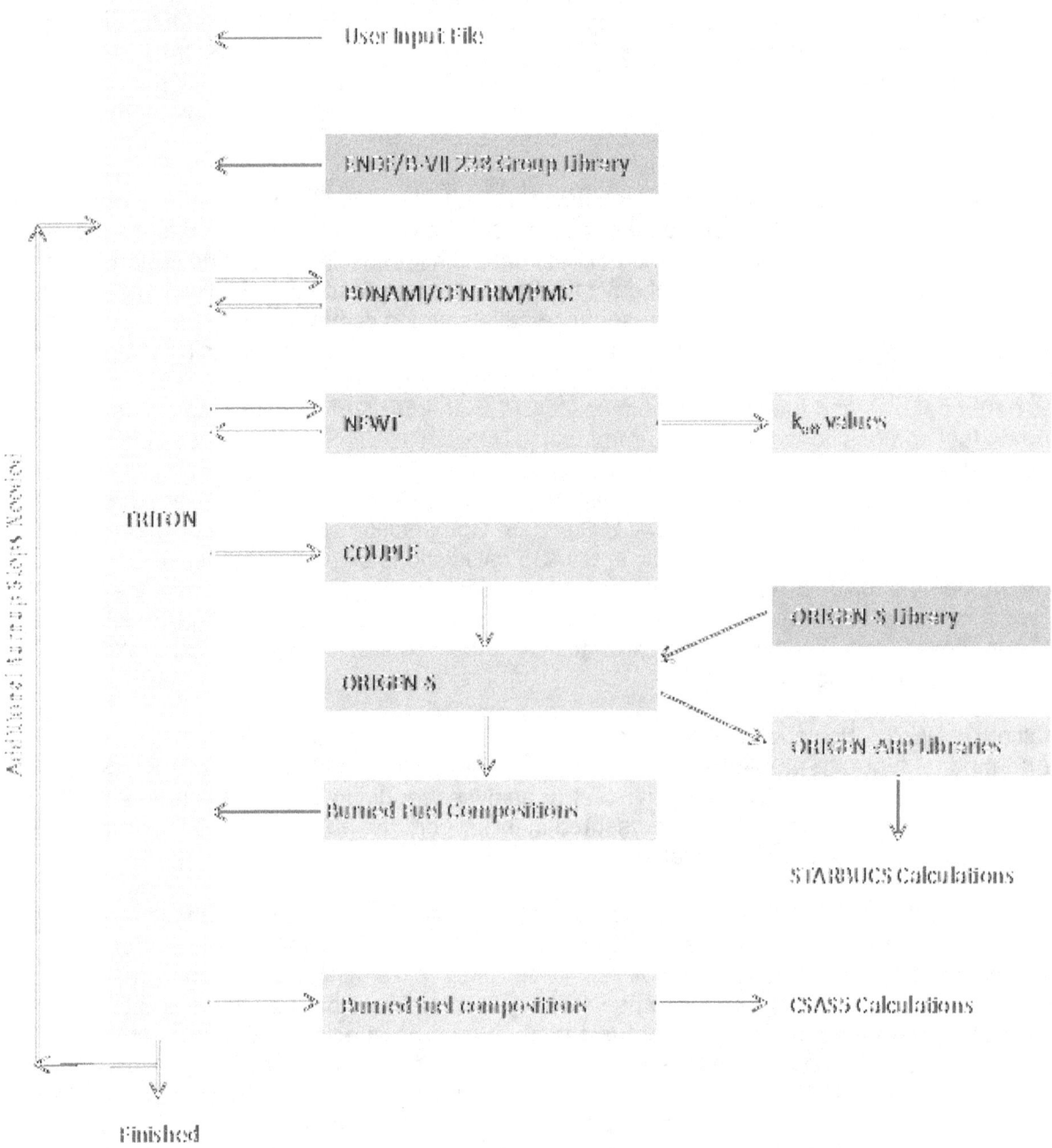

Figure 4. TRITON Calculation Program and Data Flow.

STARBUCS

The STARBUCS control module (Section C10 of the SCALE 6.1 manual), utilizes the ORIGEN-ARP depletion analysis methodology (Section D1 of the SCALE 6.1 manual), to quickly calculate burned fuel compositions using simplified inputs to describe the reactor depletion and post-irradiation cooling time. The burned fuel compositions, limited to user specified nuclides, are then inserted into a CSAS5 (KENO-V.a) or CSAS6 (KENO-VI) k_{eff} calculation. The STARBUCS calculations used in this report utilized the CSAS5 module described below.

Figure 5 is a simplified program and data flow chart for the STARBUCS sequence as used in the BWR BUC sensitivity studies. The user input file describes the geometry and materials defining the three-dimensional GBC-68 KENO model. Currently, STARBUCS is limited to starting with fresh UO_2 fuel. The user input file identifies which ORIGEN-ARP library should be used for the depletion calculations and specifies the number of days the fuel is depleted and cooled following depletion. STARBUCS passes the initial fuel composition information and ORIGEN-ARP library specification to the ARP and ORIGEN-S programs where the ORIGEN-ARP depletion analysis methodology is used to calculate burned fuel compositions, which are returned to the STARBUCS control module. STARBUCS then updates the CSAS5 input file to incorporate the burned fuel compositions and calls CSAS5.

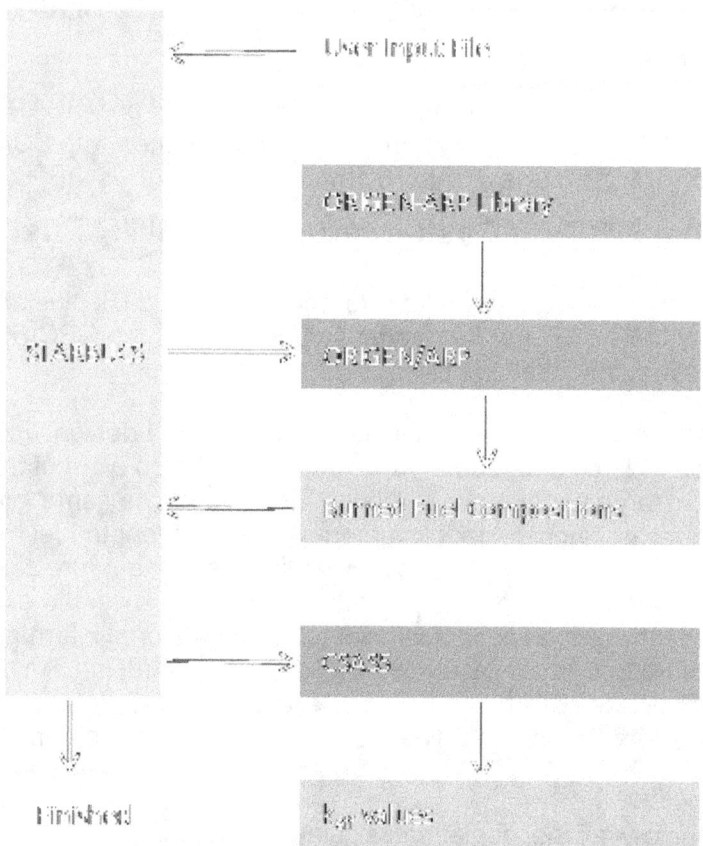

Figure 5. STARBUCS Calculation Program and Data Flow.

CSAS5

The CSAS5 control module is described in Section C05 of the SCALE 6.1 manual. It may be used with either multigroup or continuous energy nuclear cross-section data. In multigroup mode, CSAS5 performs problem-dependent resonance self-shielding calculations. CSAS5 uses the problem-dependent multigroup or the continuous energy cross-section data in a KENO-V.a Monte Carlo transport simulation to calculate the system k_{eff} value. All k_{eff} calculations performed for this report utilized the CSAS5 control module and multigroup nuclear data.

Nuclear Data

Two sets of nuclear data were used for the calculations supporting this analysis. The following description of the data used by ORIGEN-S is provided in the ORIGEN-S documentation in the SCALE 6.1 manual:

> A major upgrade of the nuclear data libraries (described in Sect. M6) has been performed for this release. The libraries include nuclear data for 2226 nuclides produced by neutron activation, fission, and decay. All decay data are based on ENDF/B-VII.0. The multigroup cross-section data are developed from the JEFF-3.0/A neutron activation file containing cross-section data for 774 target nuclides and 23 different reaction types. In addition, energy-dependent fission product yields from ENDF/B-VII.0 are included for 30 fissionable nuclides.

The neutron transport calculations performed within the TRITON and CSAS5 sequences utilized the ENDF/B-VII based 238 neutron energy group library distributed with SCALE 6.1. This library is described in Section M4.2.4 of the SCALE 6.1 manual.

3.2.2 Model of the GBC-68 Cask Loaded with BWR Fuel

The work presented in this report utilizes the reference GBC-68 cask plus spent fuel model defined in NUREG/CR-7157 [15]. For the reader's convenience, the GBC-68 cask and burned fuel assembly models are described in this section.

For the BWR sensitivity calculations, a simplified BWR assembly design was adopted. The design, similar to a GE14 lattice, utilizes a 10 × 10 lattice in which eight of the fuel rods have been replaced with two large water rods. The base model uses the same initial ^{235}U enrichment for all fuel rods and does not include gadolinium fuel rods, part-length fuel rods or axial blankets. Thorough evaluation of the impact of these simplifications is left to future studies. Fuel assembly geometry information used in the model is presented in Table 1. With the exception of the active fuel length and fuel channel dimensions, fuel assembly dimensional information was taken from Table D1.A.3 in the SCALE 6.1 manual for the GE14 assembly design. An active fuel height of 381 cm was used to facilitate use of data from LaSalle Unit 1 CRC data [17] in sensitivity studies that are planned for future work. The fuel channel dimensions from the LaSalle Unit 1 CRC data were used.

Except where noted below, all fuel depletion calculations were performed using the depletion parameters specified in Table 2. For sensitivity calculations other than the variation of moderator density during depletion, a moderator density of 0.6 g/cm^3 was used. For sensitivity calculations other than the study in Section 3.6 showing the impact of rodded versus unrodded depletion, all depletion calculations were performed with control blades fully inserted. The SCALE 6.1 CSAS5 cask and fuel assembly model described above was used for all k_{eff}

calculations, except for the sensitivity study in Section 3.10 showing the relative reactivities of the various lattices.

Table 1. BWR fuel assembly specifications[†]

Parameter	inches	cm
Fuel pellet outside diameter	0.3449	0.876
Cladding inside diameter	0.3520	0.894
Cladding outside diameter	0.4039	1.026
Cladding thickness	0.0260	0.066
Fuel rod pitch	0.5098	1.295
Water rod inside diameter	0.9138	2.321
Water rod outside diameter	0.9925	2.521
Water rod radial thickness	0.0394	0.100
Active fuel length	150.0	381.0
Array size	10 × 10	
Number of fuel rods	92	
Number of water rods	2	
Fuel channel inner dimension	5.278	13.406
Fuel channel outer dimension	5.478	13.914
Fuel channel thickness	0.100	0.2540

[†] Reproduced from Table 3 of NUREG/CR-7157

Except where specifically noted below, burned fuel composition calculations were performed using the STARBUCS sequence and ORIGEN-ARP libraries. ORIGEN-ARP libraries were generated for the 10 x 10 lattice with 8 fuel pins replaced with two large water holes. This lattice was modeled in TRITON and was depleted with the control blades fully inserted for the entire depletion, thereby providing a bounding model of control rod use. This is conservative because, while BWRs use control blades to control reactivity and power distribution more than PWRs, each assembly is exposed to control blades only a fraction of each cycle and are typically not fully inserted. Cycle-specific BWR design calculations include planned control blade insertion and withdrawal to manage core reactivity and power distribution peaking. With all other things held constant, depletion with control blades inserted results in a harder neutron energy spectrum, which increases plutonium production, thus increasing fuel reactivity. Depletion calculations were performed with TRITON to generate ORIGEN-ARP libraries for initial enrichments varying from 1.5 to 6 wt % ^{235}U, with moderator densities varying from 0.1 to 0.8 g/cm^3, for fuel axial zone average burnup values up to 88.5 GWd/MTU. The parameters used in the TRITON depletion model are provided in Table 2, and the fuel depletion model is shown in Figure 6. This table and figure were reproduced from NUREG/CR-7157. The depletion parameters used are consistent with the parameters used to generate the GE14 ORIGEN-ARP library distributed with SCALE 6.1 and are considered representative of average conditions fuel might experience during irradiation. These TRITON calculations produced ORIGEN-ARP

17

libraries that were used in STARBUCS calculations that produced the burned fuel compositions used in the BWR sensitivity study calculations presented below.

Table 2. Fuel depletion parameters

Parameter	Value
Pellet average fuel temperature (K)	840
Pellet average power density (MW/MTU)	40
Clad average temperature (K)	567
Moderator temperature (K)	512
Moderator density (g/cm^3)	
Water around fuel rods	0.1 to 0.8
Water inside water rod	0.776
Water between channels	0.776

A 512 K moderator temperature was used during the depletion calculations that generated the new ORIGEN-ARP libraries to be consistent with the conditions used to generate the GE14 ORIGEN-ARP libraries distributed with SCALE 6.1. This moderator temperature is low for a BWR operating around 1030 psia where T_{sat} is about 560 K. However, the moderator temperature has only a minor impact on the depletion calculations because the water density, which has the primary impact on the flux spectrum, is modeled explicitly and independently of the moderator temperature. The moderator temperature variation has only a minor impact through temperature-dependent adjustments to the scattering cross sections. The 840 K fuel temperature also appears low compared to some of the data presented later in Sect. 3.7, but was retained to be consistent with the value used to generate the SCALE 6.1 GE14 ORIGEN-ARP libraries.

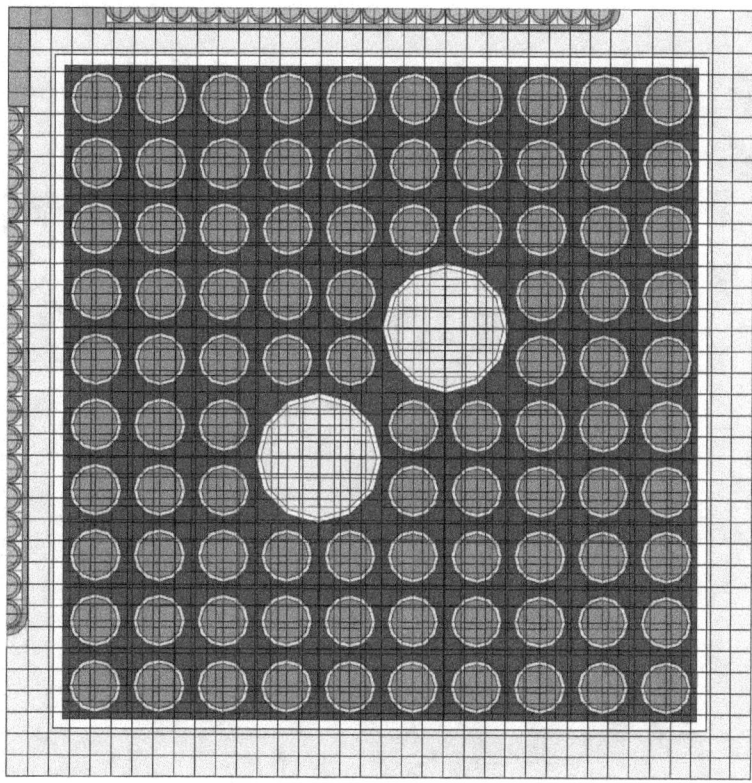

Figure 6. TRITON depletion model used to generate the rodded ORIGEN-ARP libraries.

Except where noted below, sensitivity calculations were performed using the STARBUCS sequence and the 10 x 10 assembly ORIGEN-ARP libraries that were generated specifically for this project. STARBUCS calculations were performed with the fuel assembly model described earlier in this section placed in a 68 BWR assembly generic burnup credit cask benchmark model referred to as GBC-68.

For use in some of the sensitivities studies discussed in this report, axial burnup, moderator density, and fuel temperature distributions were extracted for assembly ID B2 from the LaSalle Unit 1 CRC data [17]. This assembly was selected because it had the highest peak axial burnup of any assembly reported in the CRC data. Additional details concerning the axial distributions used are provided below in the relevant sections.

3.2.3 The GBC-68 Cask Model

The GBC-68 cask model is described in detail in NUREG/CR-7157 [15]. The cask body in the GBC-68 model is modeled as stainless steel 304 and has an inner diameter of 175 cm, a side wall thickness of 20 cm, a bottom thickness of 30 cm, and a top lid thickness of 20 cm. The cask basket holds 68 BWR fuel assemblies and is constructed such that each fuel assembly sits in a square stainless steel 304 can having an inside dimension of 15.0435 cm and a wall thickness of 0.75 cm. Storage cells have a 16.80 cm center-to-center spacing. A BORAL™ panel having a 0.020 g ^{10}B/cm^2 loading is placed between each assembly and on all cell exterior locations. Figure 7, Figure 8 and Figure 9 illustrate some of the GBC-68 cask model details. These figures were reproduced from NUREG/CR-7157.

Figure 7. Cross-sectional view of assembly cell in GBC-68 cask model.

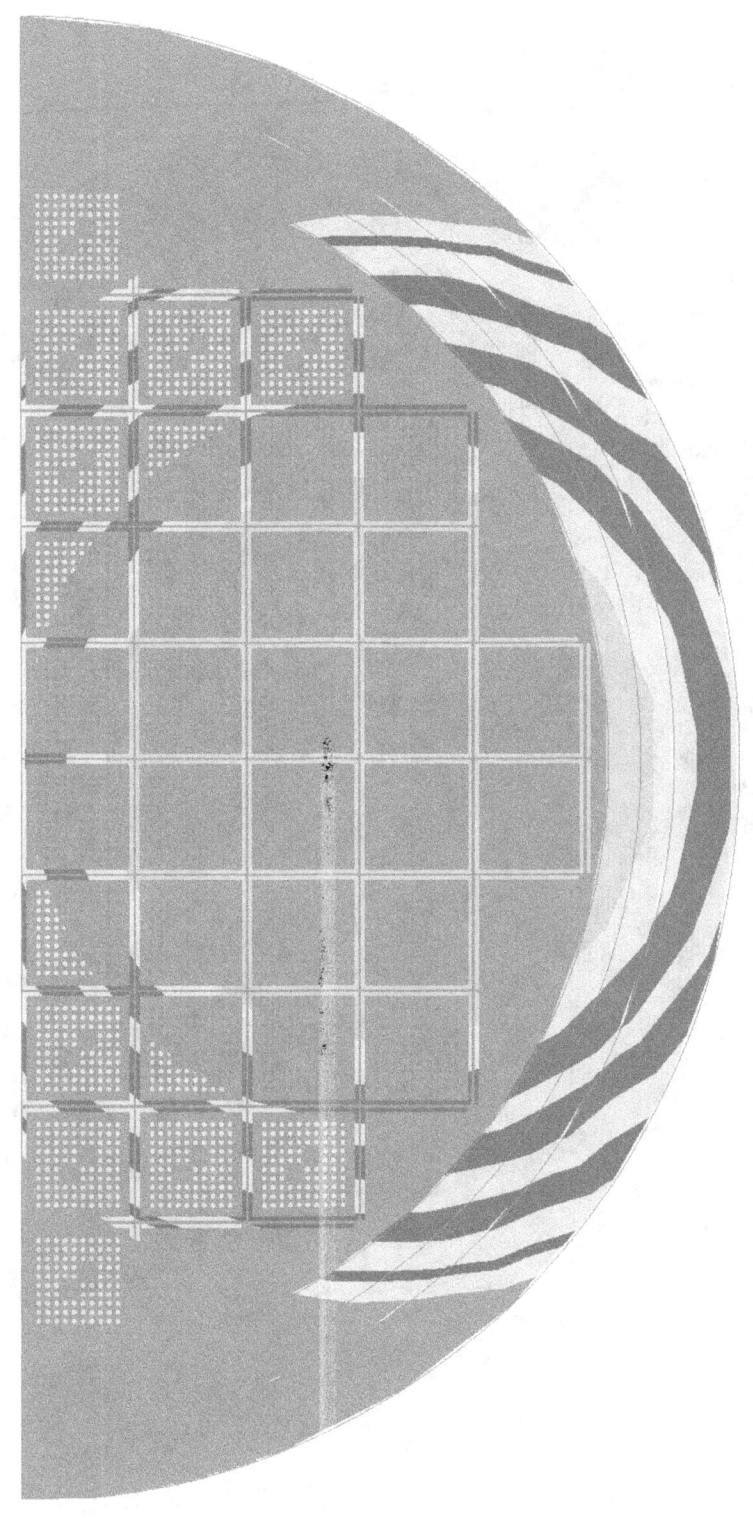

Figure 8. Radial cross section of the GBC-68 model, which uses a reflective boundary condition on the left-hand side.

21

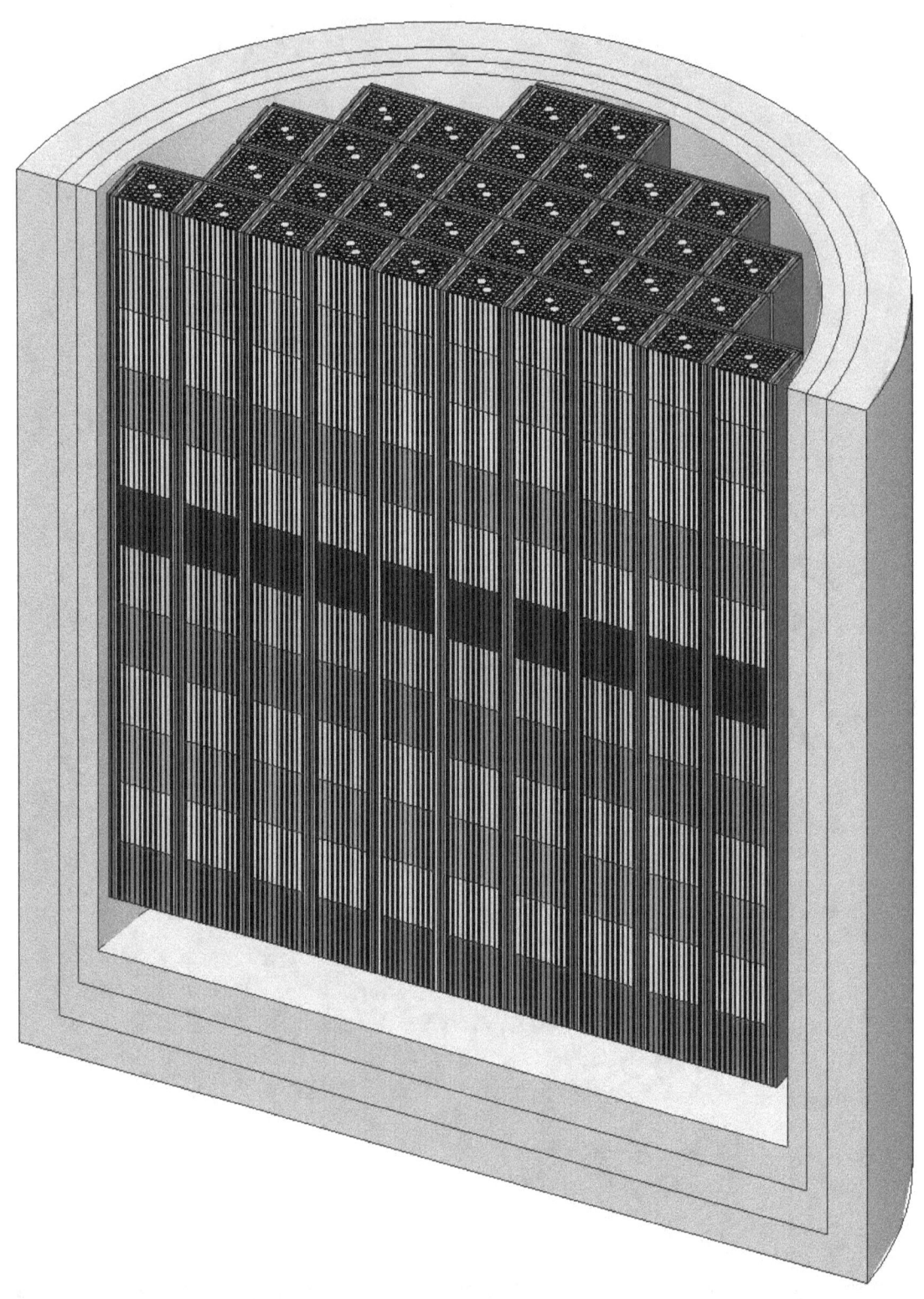

Figure 9. Cutaway view of bottom part GBC-68 model. Poison panel and cell walls were removed to reveal fuel assembly axial zoning. Note that the reference cases utilized a single axial zone.

3.3 BURNUP CREDIT NUCLIDES

An issue that needs to be revisited for BWR burnup credit is the specification of modeled burnup credit nuclides. In general, any actinide that can increase the k_{eff} of the SNF storage system should be included in the burnup credit nuclide set. The set may include radioactively stable or near-stable actinides and fission products that reduce k_{eff}. Reactivity reducing actinides or fission products that may escape the burned fuel should not be credited. Burnup credit nuclide selection should also include consideration of data available for use in validating calculations of burned fuel compositions and of k_{eff} for storage and transportation systems containing the burnup credit nuclides.

The recently issued guidance for burnup credit for PWR spent fuel in transportation and storage casks [4] recommends that credit be restricted to specified actinide and fission product compositions associated with UO_2 fuel irradiated in a PWR. Guidance documents for burnup credit for PWR fuel in SFPs do not include recommendations identifying burnup credit nuclides. SFP burnup credit analyses for PWR SFPs have credited a wider range of both actinides and fission products than has been recommended for transportation and storage casks.

For the purposes of this report, two sets of actinides and fission products are listed in Table 3. Set 1 includes nine "major actinides" that are typically modeled in actinide-only (AO) burnup credit. Set 2 includes the major actinides, three minor actinides and 16 of the most important fission products. These nuclide sets are consistent with the nuclide sets specified in ISG-8R3 [4] and used in References 15 and 22.

Table 3. Burnup Credit Nuclides

Set 1: Major actinides (9 total)								
^{234}U	^{235}U	^{238}U	^{238}Pu	^{239}Pu	^{240}Pu	^{241}Pu	^{242}Pu	^{241}Am

Set 2: Major actinides, 3 minor actinides, and 16 major fission products (28 total)								
^{234}U	^{235}U	^{238}U	^{238}Pu	^{239}Pu	^{240}Pu	^{241}Pu	^{242}Pu	^{241}Am
^{236}U	^{237}Np	^{243}Am						
^{95}Mo	^{99}Tc	^{101}Ru	^{103}Rh	^{109}Ag	^{133}Cs	^{147}Sm	^{149}Sm	
^{150}Sm	^{151}Sm	^{152}Sm	^{143}Nd	^{145}Nd	^{151}Eu	^{153}Eu	^{155}Gd	

Theoretically, the same nuclides could be credited in both PWR and BWR burnup credit. However, because there are limited radiochemical assay data available for validating fission products in BWR burned fuel compositions, it may be appropriate to limit which nuclides are credited. This issue is discussed further in Sections 3.12 and 3.13. If the gadolinia added to the Gd-fuel rods is credited in the burnup credit analysis, additional gadolinium nuclides such as ^{157}Gd could be included.

3.4 REACTOR OPERATING HISTORY

In the context of burnup credit, reactor operating history generally refers to the power level at which the reactor was operated during each cycle, the length of each cycle, and the amount of time between cycles. Reactor operating history affects the spent fuel compositions by changing the rates at which uranium is depleted, plutonium is generated, and fission products and other actinides are generated. During the shutdown time between cycles, ^{241}Pu decays to ^{241}Am and some radioactive actinides and fission products decay, while others build in with the decay of radioactive precursors.

Studies [7, 23, 24] have been performed of the impact of reactor operating history on spent fuel reactivity. These studies examined the impact of part-power operations and variation in the shutdown time between cycles and showed that the operating history variation generally had a smaller impact on spent fuel reactivity as compared to the more important parameters such as axial burnup profile, moderator density and control blade insertion, each of which may impact k_{eff} by several %Δk.

Much of the work that has been done in this area is fairly old and some was performed using simple one-dimensional depletion computer codes. The impact of BWR operating history should be revisited using modern computer codes and data.

3.5   AXIAL MODERATOR DENSITY DISTRIBUTION

By design, the boiling water reactor generates steam as the reactor coolant moves up through the fuel. As can be seen by examining the best-estimate CRC data presented in References 16, 17, and 18, the moderator density decreases from roughly 0.75 g/cm^3 at the bottom of the core to values as low as 0.2 g/cm^3 near the top of the core. This range is considerably wider than the range of moderator densities present in PWRs.

In PWR burnup credit, it is frequently assumed that the entire fuel assembly is depleted at the minimum water density for the reactor core. This results in a harder neutron spectrum, increasing the generation of plutonium, which leads to higher assembly reactivity. A similar approach could be followed for BWR burnup credit. Due to the wider range of moderator densities seen in the reactor, the extensive use of control blades for reactivity control, the use of BWR fuel channels, and the axial design features used in BWR fuel, using the lowest water density may be overly conservative. Note that the assemblies next to inserted control blades have significantly lower power levels and, consequently, have higher water densities. Using combined conservative reactor depletion conditions (i.e., control blades inserted, minimum water density, maximum fuel temperature, limiting axial-zone lattice, etc.) may be unnecessarily conservative.

Calculations were performed to illustrate the impact of the moderator density used during fuel depletion on BWR fuel reactivity in the GBC-68 cask model [15]. In each of these calculations, a single axially uniform water density was used during the depletion and a single axially uniform spent fuel composition was used for all assemblies in the GBC-68 model. Figures 10 and 11 show the variation in Δk_{eff} as a function of water density and fuel assembly burnup for fuel at two different initial enrichments, 2 and 5 wt % ^{235}U. These figures show that relatively modest changes in the effective water density used during depletion have significant effects on the calculated k_{eff} values. In general, the change in k_{eff} is up to a few percent Δk at low burnup and increases to 10%Δk at high burnups. An extensive set of results is presented in Appendix A. These data highlight the importance of using appropriate moderator density values during fuel depletion calculations. The moderator density used during depletion has a major impact on the burnup-dependent fuel reactivity. Further study in this area is obviously warranted.

The difference in the burnup dependent behavior of the data presented in Figures 10 and 11 is because the plutonium builds up significantly faster for the 2 wt % initial enrichment fuel than it does for the 5 wt % initial enrichment fuel. Some of the neutrons that would be absorbed into the ^{235}U in the 5 wt % fuel are instead absorbed in ^{238}U and other actinides in the 2 wt % fuel, many of which eventually decay to plutonium nuclides. For the 2 wt % fuel, the plutonium nuclides approach their in-reactor equilibrium values faster than in the 5 wt % fuel. It is expected that the slope of the 5 wt % fuel Δk value curves would turn over and look more like the 2 wt % fuel curves at burnups higher than 60 GWd/MTU as the plutonium nuclides reach their in-reactor equilibrium values.

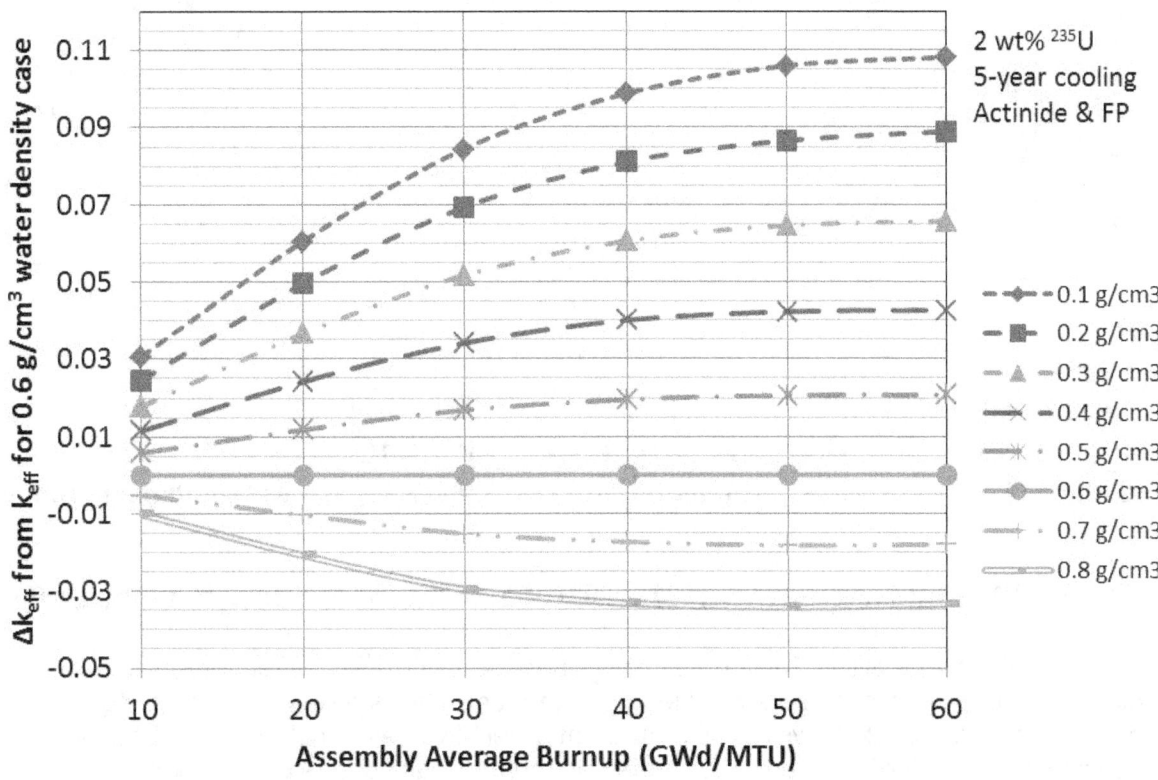

Figure 10. Change in Δk_{eff} with moderator density variation, 2 wt % initial enrichment.

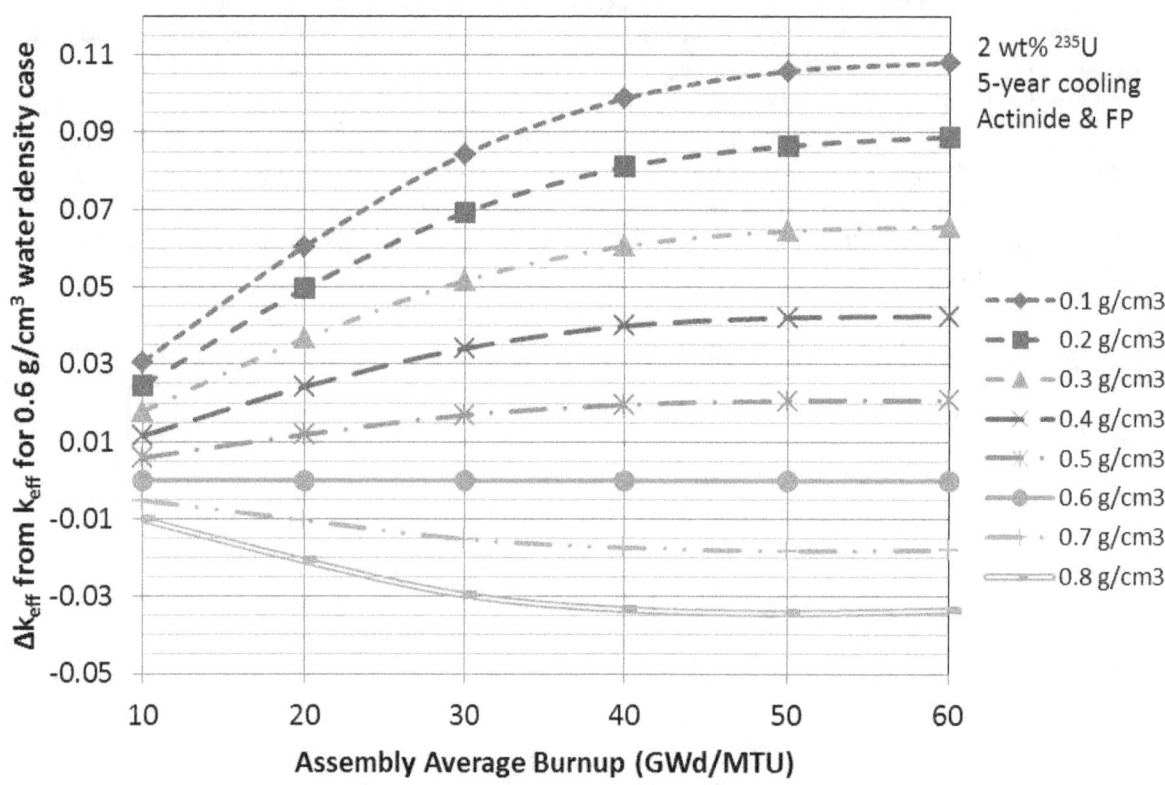

Figure 11. Change in Δk_{eff} with moderator density variation, 5 wt % initial enrichment.

3.5.1 Additional Results with Non-Uniform Axial Moderator Density

Each data point presented in Section 3.5 was generated using a single moderator density in the depletion calculation for the entire axial length of the fuel. This is a very unrealistic moderator density profile. Figure 12 shows how the moderator density in assembly B2 from LaSalle Unit 1 (data from Table 4-15 of Reference 17) varied as the fuel was used. Axially dependent data for assembly ID B2 were used in this report, because that assembly had the highest peak burnup value of all assemblies described in the LaSalle Unit 1 CRC data [17]. This assembly was used in Cycles 4, 5, 6 & 7 of LaSalle Unit 1. While the source document does not describe the source of the moderator density data, the reader should keep in mind that these data are likely inferred from a combination of reactor measurements, core nodal simulations, and core design calculations. They do provide an illustration of realistic operating conditions.

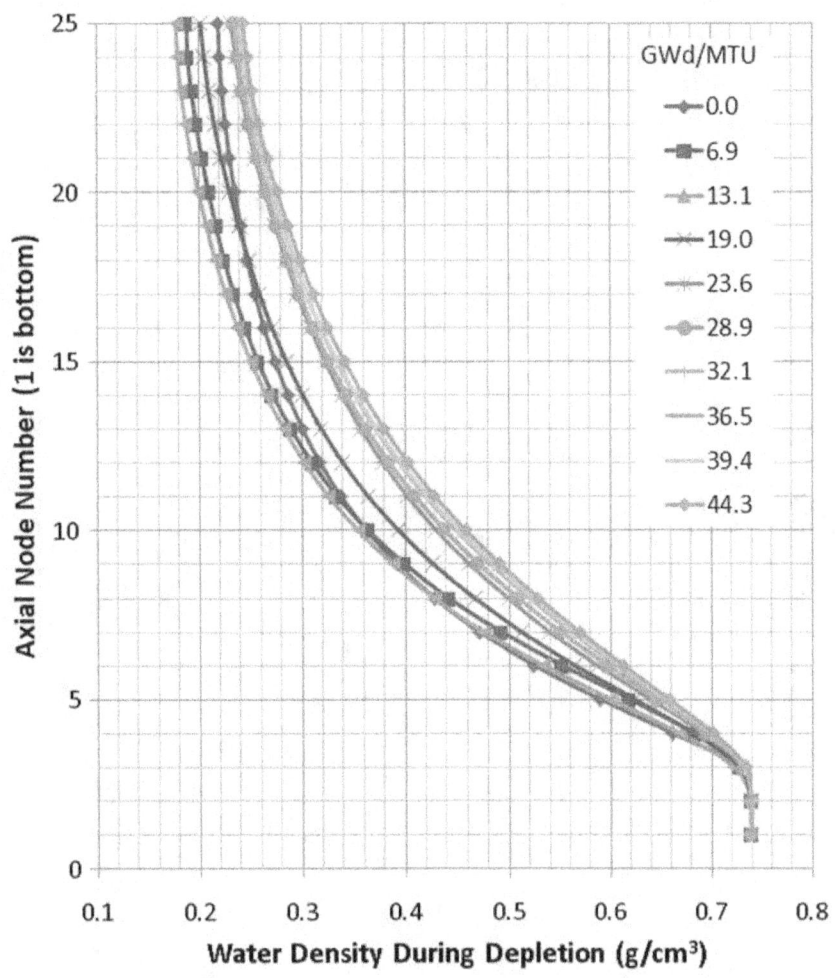

Figure 12. Effective moderator density for LaSalle Unit 1 Assembly B2 as a function of fuel assembly average burnup [17].

To evaluate the impact of axial variation of moderator density, calculations were performed using the burnup-weighted average axially dependent moderator density profile from LaSalle Unit 1 assembly ID B2 and using the assembly lifetime average moderator density for this same profile. Figure 13 shows both the burnup weighted axially-dependent profile and the assembly lifetime average moderator density value.

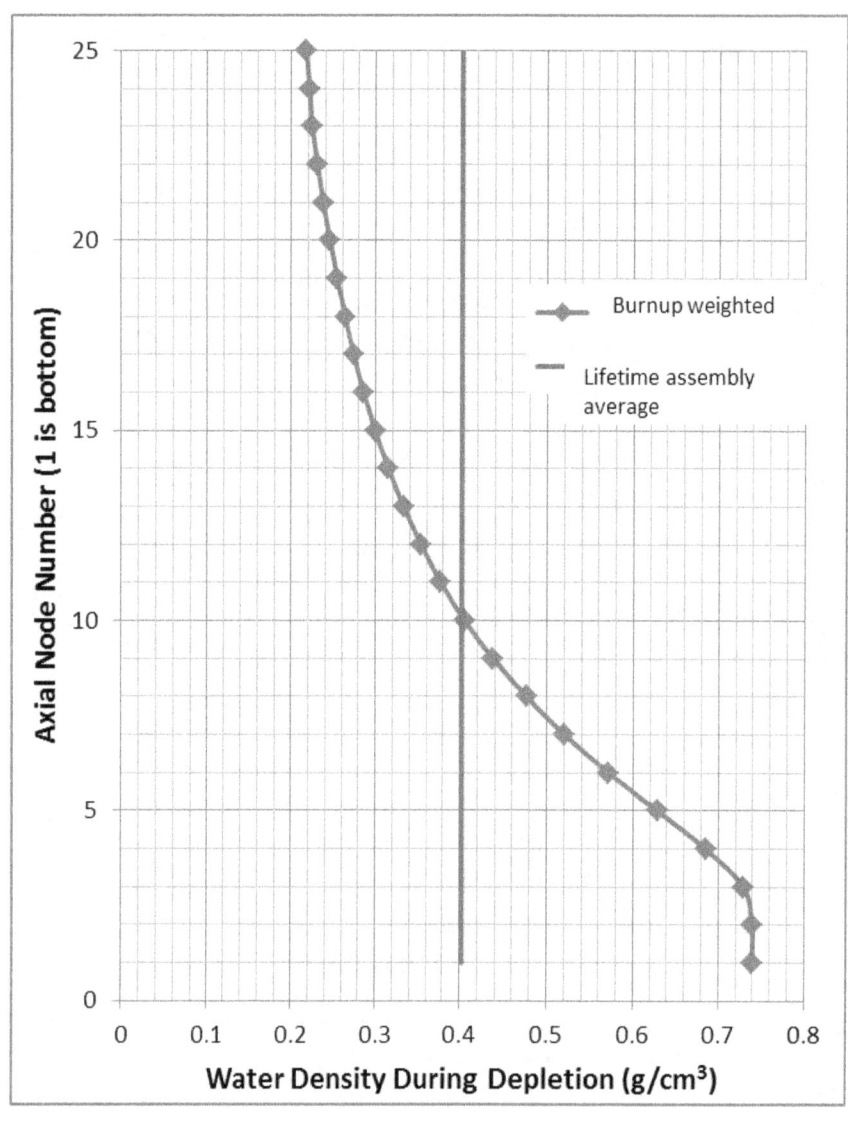

Figure 13. Burnup-weighted axially-dependent and core average moderator density profiles used during fuel depletion.

STARBUCS calculations were performed for the GBC-68 cask loaded with 10 x 10 - 8 fuel assemblies having a 4 wt % ^{235}U initial enrichment and using the axial moderator density profiles shown in Figure 13 for actinide-only (AO) and actinides plus 16 fission products (AFP) burnup credit. The change in k_{eff} associated with using axially-dependent moderator densities versus assembly average moderator density is provided in Table 4 and Figure 14. The data in Table 4 and Figure 14 show that using an assembly average value could be non-conservative by as much as 3% Δk. Note that this value is based on the estimated moderator densities for only one fuel assembly and the results indicate that this issue deserves further study. Since the results are for only one assembly, the 3% Δk value is not a usable bounding penalty to be applied to calculations using core average moderator densities. Using a single bounding low moderator density for fuel depletion calculations may be too conservative and using a single assembly average moderator density is clearly non-conservative. The impact of using axially-varying moderator densities in depletion calculations requires additional study.

Table 4. Change in k_{eff} due to use of axially-dependent moderator densities versus use of a single assembly-average moderator density

Burnup (GWd/MTU)	GBC-68 Actinide-Only BUC Δk_{eff} ($k_{axial\ distribution}$ - $k_{average}$)				
	Post-Irradiation Cooling Time (years)				
	0	5	10	20	40
10	0.0021	0.0017	0.0018	0.0016	0.0013
20	0.0055	0.0051	0.0051	0.0052	0.0047
30	0.0107	0.0105	0.0103	0.0101	0.0104
40	0.0166	0.0164	0.0165	0.0170	0.0169
50	0.0234	0.0233	0.0236	0.0235	0.0243
60	0.0293	0.0292	0.0294	0.0295	0.0303

Burnup (GWd/MTU)	GBC-68 Actinide & 16 FP BUC Δk_{eff} ($k_{axial\ distribution}$ - $k_{average}$)				
	Post-Irradiation Cooling Time (years)				
	0	5	10	20	40
10	0.0016	0.0013	0.0014	0.0012	0.0011
20	0.0050	0.0048	0.0047	0.0047	0.0044
30	0.0099	0.0101	0.0101	0.0102	0.0100
40	0.0159	0.0163	0.0163	0.0168	0.0170
50	0.0226	0.0230	0.0235	0.0232	0.0239
60	0.0282	0.0286	0.0291	0.0293	0.0295

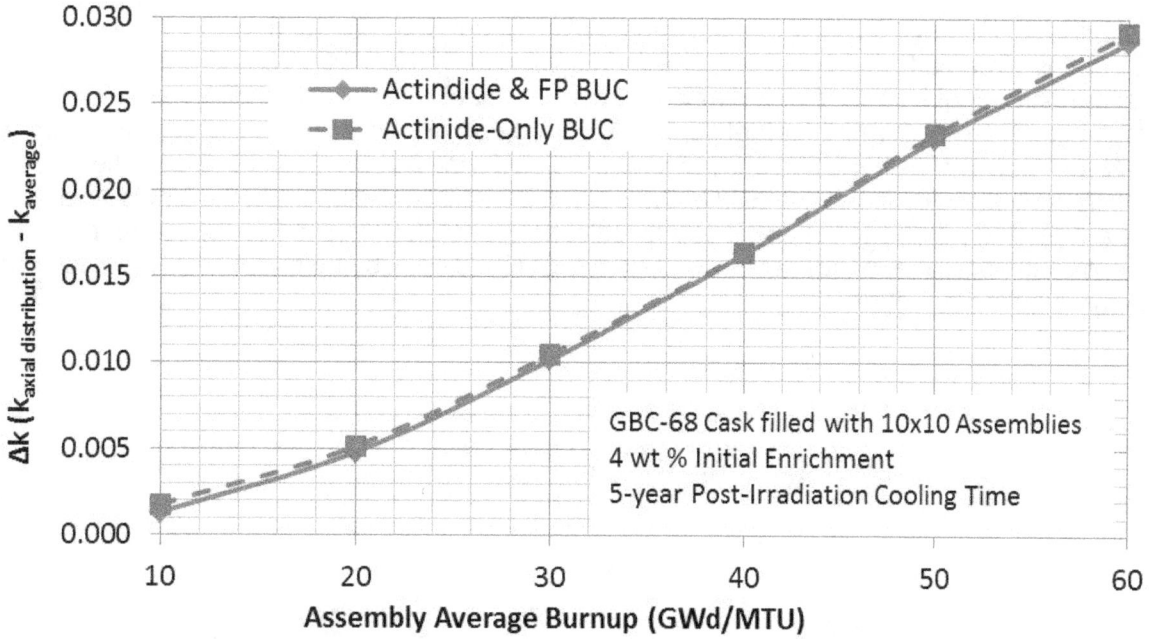

Figure 14. Change in k_{eff} due to use of axially-dependent moderator densities versus use of a single assembly-average moderator density.

One issue not addressed in this study is the sensitivity of k_{eff} to changes in the moderator density for the bypass flow, which includes the water in the water rods and outside the fuel channels. It is expected that realistic variations in the bypass flow would have minor impacts on the burned fuel compositions and the resulting k_{eff} values generated. Note that the ORIGEN-ARP libraries used for this report used a bypass flow density of 0.776 g/cm^3. This is the value that was used in the generation of the GE14 ORIGEN-ARP library distributed with SCALE 6.1 and documented in Table D1.A.3 of the SCALE 6.1 manual.

The sensitivity study presented in this section evaluated the impact of varying the "effective" moderator density, which includes the effect of steam voids. In reality, the effective moderator density is a function of local water temperature, local void fraction, and the reactor pressure. The calculations performed for the sensitivity study focused solely on the effective moderator and did not model moderator temperature variation. Thus, the nuclear data used by the depletion code in the sensitivity calculations were based on the temperatures provided in Table 2 and did not include any temperature related adjustments for the variation in moderator temperature. Future research related to the impact of moderator properties on burned fuel composition calculations could include assessment of the impact of moderator temperature along with moderator density.

3.6 CONTROL BLADE USAGE

An important difference between PWR and BWR operations is that BWRs use control blades extensively for power distribution and reactivity control, while PWRs use control rods only rarely at full power. BWR control blades are inserted from the bottom and PWR control rods are inserted from the top. Typically, only a small fraction of control blades are inserted at full power. Which control blades are inserted, the depth of insertion and length of time of insertion vary. Consequently, it is very unlikely that any assembly will be depleted with a control blade completely inserted adjacent to the assembly throughout its depletion.

To provide an estimate for the potential impact of depletion with control blades inserted, calculations were performed for the GBC-68 cask loaded with the 10 x 10 assemblies having an initial enrichment of 4 wt % ^{235}U and having a post-irradiation cooling time of 5 years. Calculations were performed for fuel depleted with control blades fully inserted and with control blades withdrawn. The depletion calculations were performed at the conditions specified in Table 2 and with a moderator density of 0.6 g/cm^3. The GBC-68 model used a single axially constant burned fuel composition. These calculations show that depletion with control blades present could increase reactivity of the fuel in the GBC-68 cask by up to 8% Δk. The results are presented below in Figure 15.

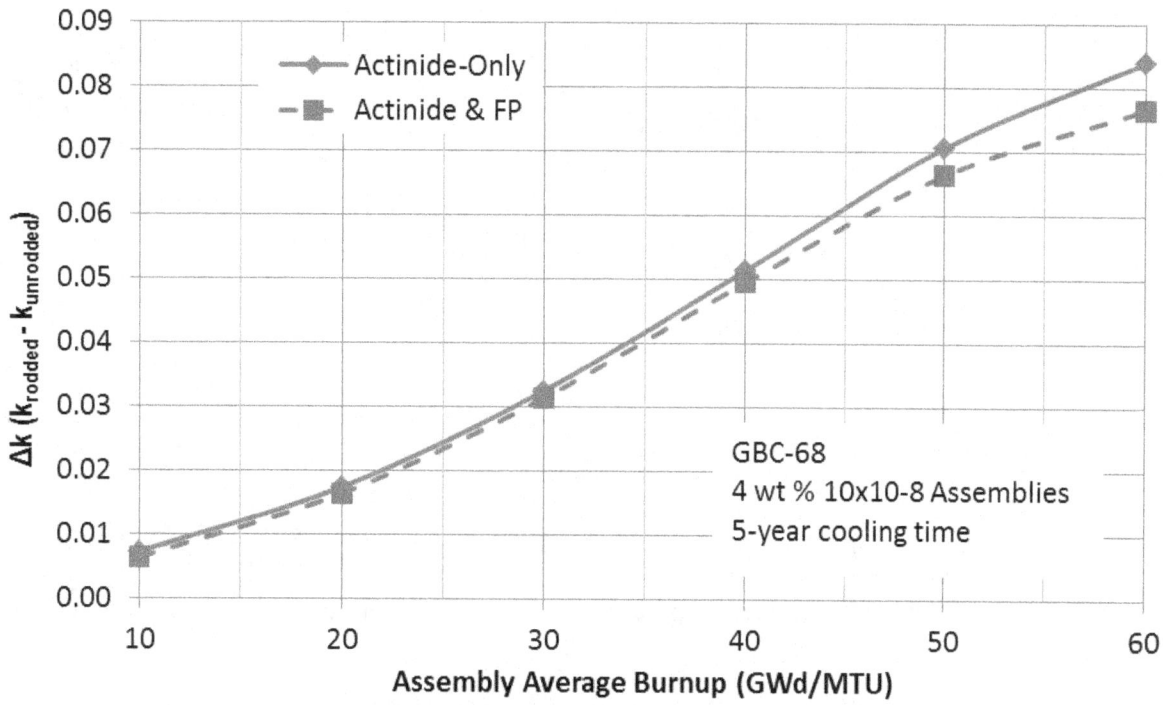

Figure 15. Change in k$_{eff}$ resulting from depletion with control blades present.

Further studies of the effects associated with fuel depletion with control blades present should consider the impact of partial insertion, the duration of insertion, and the interrelation between control blade insertion, moderator density, fuel temperature, and axial burnup distribution.

3.7 FUEL TEMPERATURE

Increasing the fuel temperature during depletion increases the reactivity of the fuel in the cask storage environment through Doppler broadening of ^{238}U resonances in the reactor, which results in increased absorption of neutrons by ^{238}U and thus increased plutonium generation. In PWR burnup credit, fuel temperatures are generally handled by using a single bounding fuel temperature, frequently 1000 or 1100 K. Selection of a bounding temperature for BWRs is complicated by the broader range of heat transfer conditions experienced by the core. Figure 16 shows the estimated fuel temperatures for assembly B2 from the LaSalle Unit 1 CRC data [17] as a function of axial node for several assembly average burnups. Some of the axial features apparent in the curves in Figure 16 are due to axial design features built into the assembly. Note that, for this assembly, fuel temperatures are significantly higher at lower burnups. This is due, in part, to a higher power density at lower burnups. Based on this data, use of a 1200 K fuel temperature might be appropriate. A less conservative approach might be to use 1200 K up to an assembly average burnup of 19 GWd/MTU and 1000 K beyond 19 GWd/MTU. As was noted in Section 2.2, the sources for and uncertainties associated with the fuel temperatures provided in the CRC data are unclear. The CRC data were likely inferred from a combination of reactor measurements, reactor simulations, and design calculations. Criticality analysis should utilize accurate data and include consideration of the uncertainties associated with the data.

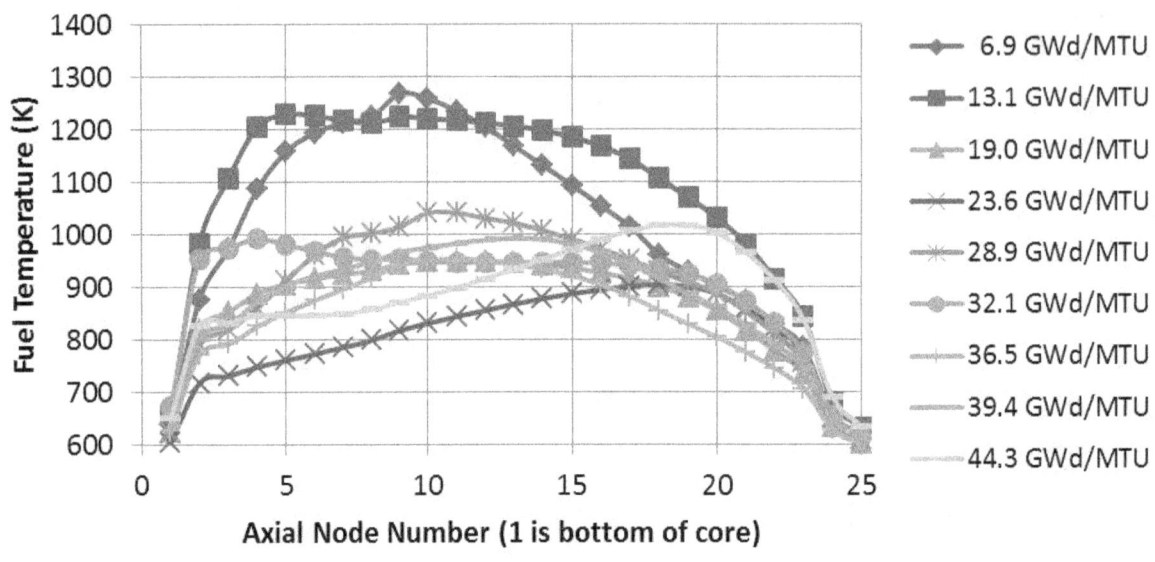

Figure 16. Burnup-dependent fuel temperatures for LaSalle Unit 1 assembly B2.

Calculations were performed to demonstrate the sensitivity of k_{eff} in the GBC-68 cask due to fuel temperature variation. The STARBUCS sequence could not be used for this series of calculations because the fuel temperature feedback during depletion is included directly in the ORIGEN-ARP libraries and cannot be adjusted. Consequently, the TRITON *T-DEPL* sequence was used to calculate the fuel compositions directly, using the same model used to generate the ORIGEN-ARP libraries, but with modified fuel temperatures. Temperature-dependent fuel compositions were then manually added to the GBC-68 models and k_{eff} calculated using the SCALE 6.1 CSAS5 sequence. Each GBC-68 k_{eff} calculation used a single axially constant burned fuel composition.   Figure 17 and   Figure 18 show the impact on k_{eff} of the change in fuel temperature as a function of fuel assembly average burnup. The 840 K reference temperature used to construct these figures is the fuel temperature used to generate the ORIGEN-ARP libraries used for STARBUCS calculation results presented in this report and is lower than the average temperatures in Figure 16. The impact of fuel temperature modeling should be revisited using modern operating data.

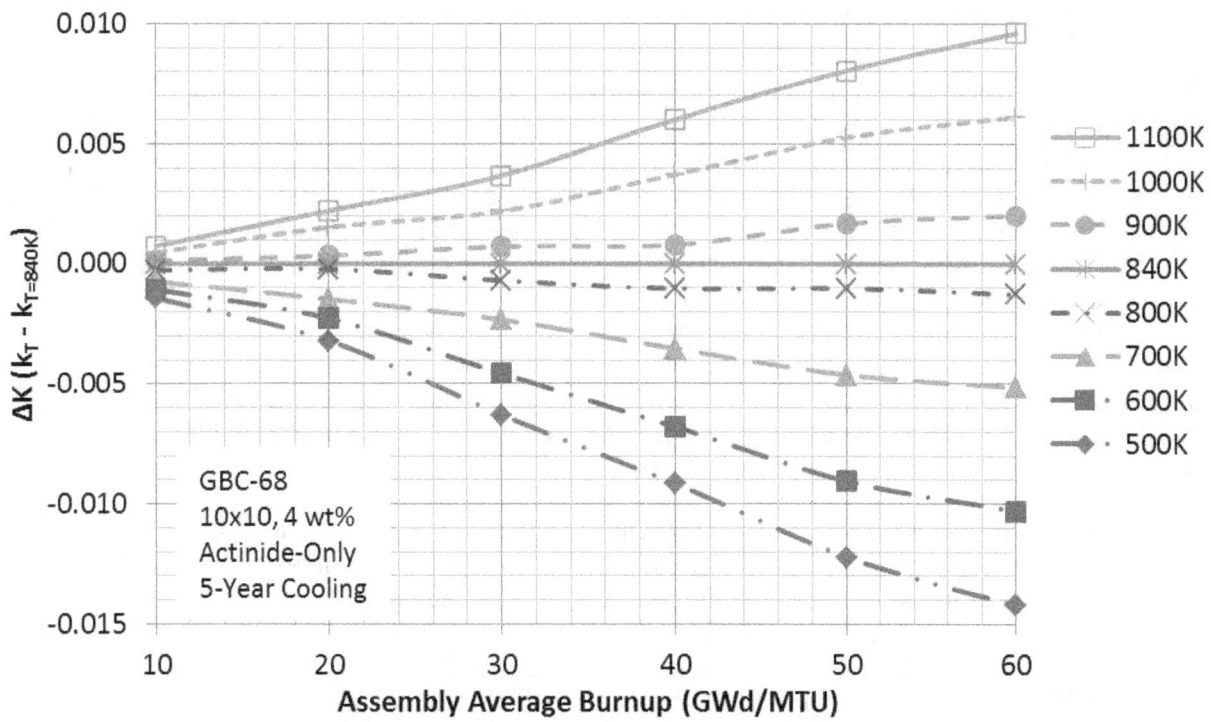

Figure 17. Change in k_{eff} from reference case (T=840K) due to fuel temperature variation for actinide-only burnup credit.

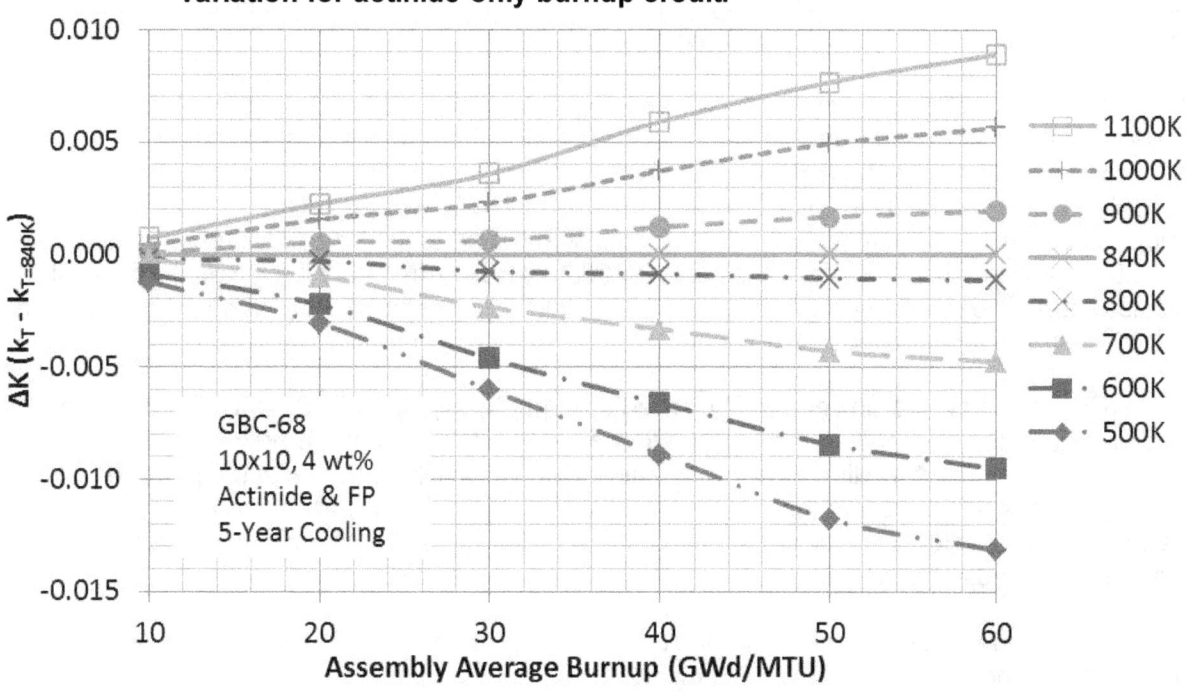

Figure 18. Change in k_{eff} from reference case (T=840K) due to fuel temperature variation for actinide + 16 FP burnup credit.

3.8 POWER DENSITY

Fuel depletion calculations utilize the power density or specific power, typically in units of watts per gram (W/g) of initial uranium or megawatts per initial metric ton of uranium (MW/MTU), to determine the rate at which fuel is burned. In the reactor, power density affects fuel depletion primarily in two ways. All other things held constant, a higher power density results in higher fuel rod temperatures. In most fuel depletion calculations, the fuel temperature is input independent of power density and, consequently, increasing power density does not increase fuel temperature. Power density also affects the rate at which fuel assembly burnup is accumulated. At a lower power density it takes more days to achieve the same assembly burnup. This affects fuel composition calculations by changing the amount of time for radioactive decay of burnup credit nuclides and their precursors. Some computer codes, such as SCALE/ORIGEN-S, can use the power density to calculate the neutron fluxes that are used to calculate the fuel compositions.

Calculations were performed to demonstrate the sensitivity of k_{eff} for fuel in the GBC-68 cask due to power density variation during irradiation. The STARBUCS sequence could not be used for this series of calculations because part of the power density effect during depletion is included directly in the ORIGEN-ARP libraries and cannot be adjusted. Consequently, TRITON was used to calculate the fuel compositions directly, using the same model used to generate the ORIGEN-ARP libraries, but with modified power densities. Power-density-dependent fuel compositions were then manually added to the GBC-68 models and k_{eff} calculated using the SCALE 6.1 CSAS5 sequence. Each GBC-68 model used a single axially-uniform burned fuel composition. Table 5 shows how changing the power density affects the calculated k_{eff} for the GBC-68 as a function of fuel assembly average burnup.

The 40 MW/MTU reference power density used to construct the Δk parts of the table is the power density used to generate the GE14 ORIGEN-ARP libraries distributed with SCALE 6.1 and to generate the 10 x 10 ORIGEN-ARP libraries used for STARBUCS calculation results presented in this report. The calculations reported in Table 5 were generated using the simplified 10 x 10 assembly model with fuel initially enriched to 4 wt % ^{235}U, and a 5-year post-irradiation cooling time. The impact of axial variation of power density on k_{eff} was not simulated for this report. The results show that the sensitivity of k_{eff} to power density is much smaller than the other parameters studied in this report.

Figure 19 shows how the power density varied throughout the life of LaSalle Unit 1 assembly B2. This figure shows that the magnitude and the axial distribution of the power density vary significantly throughout the life of each assembly. The variability of the axial shape of the power density profile is much greater than is seen in PWRs. This is due in part to the use of fuel channels in the reactor. This allows the coexistence of significantly different moderator density axial profiles in adjacent assemblies.

The impact of power density modeling, including axial variation, should be revisited using modern operating data.

Table 5. Sensitivity of k_{eff} and Δk to power density variation

	GBC-68 Actinide-Only BUC k_{eff}				
Burnup	Depletion Power Density (MW/MTU)				
(GWd/MTU)	20	30	40	50	60
10	0.9042	0.9044	0.9048	0.9049	0.9052
20	0.8546	0.8552	0.8561	0.8564	0.8566
30	0.8028	0.8040	0.8055	0.8062	0.8066
40	0.7503	0.7528	0.7550	0.7559	0.7567
50	0.7024	0.7058	0.7085	0.7099	0.7114
60	0.6625	0.6663	0.6698	0.6716	0.6732

	Δk (k_{pd} - $k_{pd=40}$)				
Burnup	Depletion Power Density (MW/MTU)				
(GWd/MTU)	20	30	40	50	60
10	-0.0007	-0.0004	0.0000	0.0001	0.0004
20	-0.0015	-0.0009	0.0000	0.0003	0.0006
30	-0.0028	-0.0016	0.0000	0.0006	0.0011
40	-0.0047	-0.0022	0.0000	0.0009	0.0018
50	-0.0060	-0.0027	0.0000	0.0014	0.0029
60	-0.0074	-0.0035	0.0000	0.0018	0.0033

	GBC-68 Actinide & FP BUC k_{eff}				
Burnup	Depletion Power Density (MW/MTU)				
(GWd/MTU)	20	30	40	50	60
10	0.8736	0.8725	0.8723	0.8718	0.8711
20	0.8076	0.8066	0.8064	0.8059	0.8052
30	0.7416	0.7415	0.7416	0.7407	0.7404
40	0.6777	0.6782	0.6790	0.6787	0.6784
50	0.6210	0.6221	0.6235	0.6234	0.6235
60	0.5754	0.5769	0.5786	0.5787	0.5792

	Δk (k_{pd} - $k_{pd=40}$)				
Burnup	Depletion Power Density (MW/MTU)				
(GWd/MTU)	20	30	40	50	60
10	0.0013	0.0003	0.0000	-0.0005	-0.0012
20	0.0012	0.0002	0.0000	-0.0005	-0.0012
30	-0.0001	-0.0002	0.0000	-0.0009	-0.0013
40	-0.0012	-0.0008	0.0000	-0.0002	-0.0005
50	-0.0025	-0.0015	0.0000	-0.0002	0.0000
60	-0.0033	-0.0017	0.0000	0.0001	0.0006

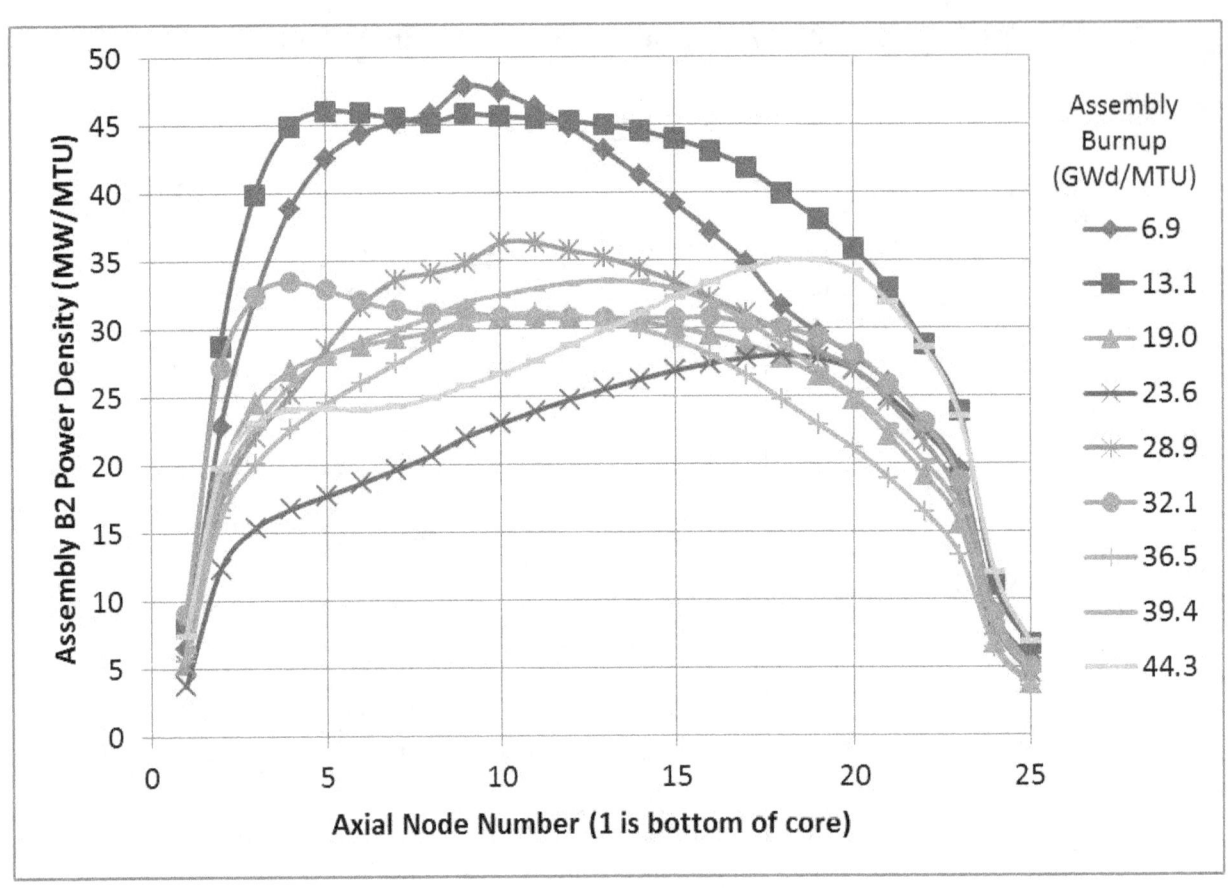

Figure 19. Power density as a function of burnup for LaSalle Unit 1 assembly B2.

3.9 MODELING OF FUEL ASSEMBLY BURNUP

As fuel assemblies are used in the reactors they accumulate burnup, which is generally quantified as the amount of energy extracted from the fuel per initial mass of uranium. The most frequently used units are megawatt days per initial metric ton of uranium (MWd/MTU) or gigawatt days per initial metric ton of uranium (GWd/MTU). Burnup is the power density (MW/MTU) integrated over time. Due to assembly design and reactor operating and control characteristics of BWRs, the power density distribution in any one assembly varies significantly more than it does in PWR assemblies. To further complicate the issue, many modern BWR fuel assemblies include part-length fuel rods. For those assemblies, the upper parts of the assembly have a relatively lower uranium density. The assembly average burnup is no longer the simple average of the burnup of the assembly regions. Instead, it is the initial MTU weighted average of the burnup from the different regions.

To date, the issue of BWR fuel burnup distribution has been avoided by restricting burnup credit to the burnup where the peak reactivity occurs for the limiting assembly with fuel rods containing gadolinia. For PWR assemblies, ignoring axial burnup distribution is typically conservative at assembly average burnups below 10 to 20 GWd/MTU. It has been implicitly assumed by BWR SFP analysts that the same would hold true for BWR assemblies. Considering the variability of the axial power density profile illustrated in Figure 19, this assumption should be checked.

Further, any burnup credit beyond peak reactivity must include an evaluation of the impact of axial burnup distribution on the reactivity of the stored fuel.

A significant amount of work has been directed toward collecting and evaluating axial burnup distributionS for PWR fuel. There are too many reports and contributors to provide a thorough review of the work here. However, a good starting point is the list of references provided in Reference 20. Much of the modern analysis associated with evaluation of axial burnup distribution for PWRs relies on an axial burnup profile database published by Cacciapouti and Van Volkinburg [25] and an evaluation of this data published by Parish and Chen [26]. Unfortunately, the axial burnup profile database contains data for only PWRs. A limited set of axial burnup profiles could be extracted from the publicly available BWR CRC data [16, 17, 18]. The results presented in NUREG/CR-6801 [20] indicate that, for PWR BUC, modeling axial burnup distribution increases k_{eff} by up to 0.08 Δk at an assembly average burnup of 50 GWd/MTU. There is little reason to expect the effect to be smaller for BWR BUC.

While some work [7] has been performed to evaluate the impact of axial burnup distribution on BWR fuel reactivity, additional study is needed to identify sources of axial burnup distribution data, quantify the impact of axial design features on axial burnup distributions, to evaluate application of the axial burnup distribution data to BWR burnup credit analyses, and to confirm the commonly used assumption that axial burnup distribution may be safely ignored for criticality analyses using the SCCG peak k_∞ method.

3.10 FUEL ASSEMBLY DESIGN VARIATION

A common practice in PWR burnup credit criticality analysis is the identification of one or more bounding fuel assembly designs that are then used in the analysis to calculate limits. Designs are bounding in the sense that for enrichments and assembly average burnup values along the authorized loading curve, the bounding design or designs produce the highest k_{eff} values. For some analyses, multiple bounding fuel assembly designs are required because the most reactive design may change with burnup.   Figure 20 provides a burnup-dependent comparison of the reactivity of 7 x 7, 8 x 8, and 9 x 9 lattices compared to a 10 x 10 lattice. The results shown are for 4 wt % ^{235}U unpoisoned assemblies, actinide-only burnup credit, for fuel with a 5-year post-irradiation cooling time. The 7 x 7, 8 x 8 and 9 x 9 lattice fuel depletion calculations used the same reactor parameters that were used for the 10 x 10 lattice and are described in Table 2. The assembly design for each lattice was taken from Table D1.A.3 of the SCALE 6.1 manual. The 7 x 7 array is identified as the GE3B assembly. The 8 x 8 array is the GE9 assembly design, and the 9 x 9 is the GE11 assembly design. For all three assembly designs the Gd fuel rods have been replaced with unpoisoned fuel rods. TRITON calculations were performed to simulate depletion of the fuel with control blades fully inserted. The resulting burned fuel compositions were then used in the standard GBC-68 cask model loaded with each of the fuel types.

Note from   Figure 20 that for these simplified models the 10 x 10 lattice is bounding for burnups below 40 GWd/MTU and the 7 x 7 lattice becomes bounding for burnups above 40 GWd/MTU. Multiple bounding designs may also be needed to support use of multiple design-dependent loading curves.

Modern BWR fuel assembly designs are more complex than PWR fuel assembly designs. BWR fuel lattices have varied from 6 x 6 to 11 x 11 fuel pin bundles, with multiple lattice sizes used at many plants. The fuel bundle reactivity will vary with lattice size, fuel pin pitch, fuel pellet density and diameter, fuel rod clad outer diameter, the number, size, and location of water rods, and the number, location and Gd_2O_3 content of gadolinium bearing fuel rods. BWR fuel assemblies have axially varying features such as natural or enriched uranium blankets, axial zoning of fuel pin uranium enrichments, axial zoning of the gadolinia concentration in the gadolinium fuel rods, and the use of part-length fuel rods, which leave vacant water-filled array positions in the upper reaches of the assemblies. BWR assemblies may also have axially and radially varying initial ^{235}U enrichments.

Studies are needed to explore the influence of the BWR fuel assembly design variations on burnup credit criticality analyses. Work should be done to explore the impact of modeling simplifications that may be used such as using maximum lattice average enrichment rather than the pin-by-pin enrichment distributions, modeling Gd fuel rods as non-Gd fuel rods, use of the limiting axial lattice for the entire assembly, storage of fuel with or without fuel channels, etc.

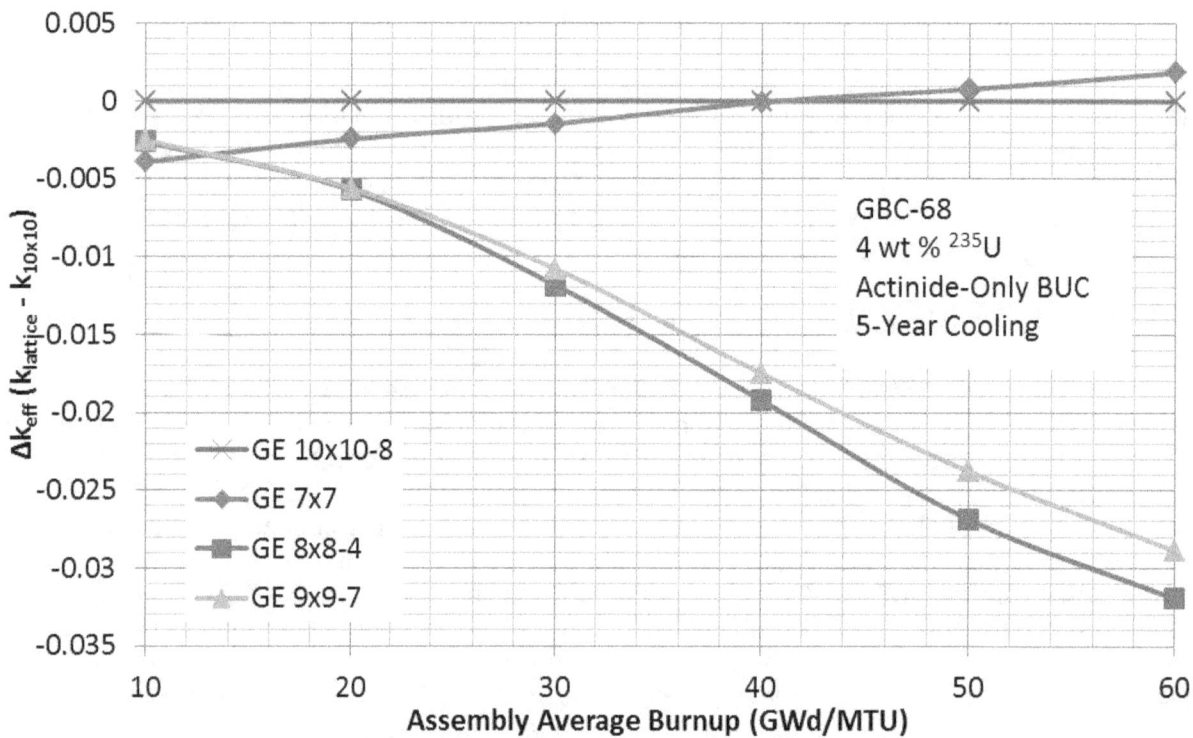

Figure 20. Burnup-dependent comparison of reactivity for various BWR lattices.

3.11 BURNUP CREDIT IMPLEMENTATION CONSIDERATIONS

Currently, BWR fuel storage analyses are based on the bounding simplification that the cask or spent fuel pool is filled with fuel assemblies that are all at the maximum reactivity point. In reality, this maximum reactivity occurs at a fairly low assembly average burnup, typically between 10 and 20 GWd/MTU. Most fuel assemblies have burnups significantly higher than the burnup where the limiting peak reactivities occur. Implementation of full burnup credit will reduce what had been previously unquantified and uncredited margin.

Due to reduced margins in both normal and accident conditions, it will be appropriate to more carefully examine some aspects of BWR burned fuel criticality analyses. For example, the fuel assemblies change physically during irradiation. The impact of fuel assembly changes, such as

38

fuel stack height growth, clad thinning, and assembly twist and bow on spent fuel storage rack or cask k_{eff} should be evaluated.

Implementation of BWR burnup credit will require additional study in some areas. For example, it will be necessary to quantify the uncertainty in the assembly average burnup used to determine acceptability for storage. If assembly burnups are inferred from in-reactor measurements, assembly burnup uncertainties will include detector measurement uncertainty, uncertainty in the translation of the detector measurements to power, uncertainty associated with inferring power in non-instrumented locations, uncertainty in the core average power level, and uncertainty in the integration of power over time. If design prediction burnup values are used, it will be necessary to quantify the bias and uncertainty between design predictions and actual assembly burnup values. While assembly burnup assignment has been examined for PWR burnup credit [27], such uncertainty analysis has not been needed for BWR fuel storage because fuel storage acceptability is based on the value of the peak k_{eff} value, without consideration of assembly burnup.

Implementation of full burnup credit will create the possibility for new fuel assembly storage errors. For PWR burnup credit in spent fuel storage pools, increased k_{eff} values related to the occurrence of misloaded fuel assemblies is easily covered by soluble boron credit, which does not exist in BWRs. Implementation of BWR burnup credit for transportation and storage casks will likely require a BWR-specific misload analysis.

An additional difference between PWR and BWR SFP burnup credit is that, due to no soluble boron credit, the BWR burnup credit analysis will need to include margin to the regulatory limit to cover accident conditions. For example, the analysis of misloading a low-burnup BWR assembly into a burnup credit fuel storage rack location will affect the specification of the loading curve.

3.12 VALIDATION OF BWR BURNUP CREDIT CALCULATIONS

Validation establishes the relationship between calculation results and reality. Burnup credit involves two distinct types of problems that need to be validated. First, a computational method is used to calculate burned fuel compositions. The calculated burned fuel compositions are then used to calculate the k_{eff} values for the safety analysis models. Historically, these methods have been validated separately, but using an appropriate validation method, it may be possible to perform an integrated validation instead.

Fuel composition calculations are typically validated by comparing calculated compositions to measured radiochemical assay (RCA) data from actual used nuclear fuel. Some validation methods that could be used are described in NUREG/CR-6811 [28] and NUREG/CR-7108 [29]. Some additional information on the analysis of the available BWR RCA data is provided in ORNL/TM-2010/286 [30]. Unfortunately, there are only 32 publicly available BWR RCA samples and these are for older designs that do not include some of the complexities present in modern BWR fuel. These RCA samples also do not contain measurement data for many of the important fission products. It may be possible through application of sensitivity and uncertainty analysis techniques to show that it is appropriate to use the PWR RCA sample data to validate BWR fuel composition calculations. This would greatly increase the number of RCA sample data points usable for BWR composition validation and the validation coverage of the important fission product nuclides. Additional work should be conducted to evaluate the potential for using PWR RCA data to validate BWR fuel composition calculations.

The work applicable to BWR burnup credit that was published in NUREG/CR-7108 was limited to validation of fuel compositions at the peak reactivity burnup of a BWR fuel assembly containing Gd fuel rods. The method demonstrated in NUREG/CR-7108 could be expanded to validate BWR burned fuel composition calculations beyond peak reactivity for BWR burnup credit applications.

It is also necessary to validate BWR burned fuel k_{eff} calculations. This is usually accomplished by modeling critical experiments that are similar to the safety analysis models. Statistical analysis is then performed on the critical experiment results to determine the bias and bias uncertainty associated with the computational method. Validation of calculated k_{eff} values is described in NUREG/CR-6698 [31] and, more specifically for burnup credit, in NUREG/CR-7109 [32]. The work applicable to BWR burnup credit that was published in NUREG/CR-7109 was limited to validation at the peak reactivity burnup of a BWR fuel assembly containing Gd fuel rods. Additional work is needed to support expansion of the BWR validation method demonstrated in NUREG/CR-7109 to k_{eff} calculations for BWR burnup credit applications with assembly burnups beyond the peak reactivity point.

3.13 DATA NEEDED TO SUPPORT FULL BWR BURNUP CREDIT

Due to the much broader range of conditions observed in operating BWRs, use of multiple potentially inconsistent bounding conditions may be unnecessarily conservative. It is unrealistic to assume that, for fuel composition calculations, the adjacent control blades are inserted, the power density is high, the fuel temperatures are high and the effective water density is low, all at the same time. Control blade insertion will result in lower thermal neutron flux, lower assembly power, higher water density, and lower fuel temperatures. Analysis of BWR operational data could be used to establish reasonably conservative fuel depletion conditions for use in burnup credit. Some BWR operating data is available in the form of CRCs [16, 17, and 18]. Additional detailed operating data could be used to strengthen the bases for BWR burnup credit.

Very little work has been done to evaluate the impact of the axial burnup distribution on BWR fuel assembly reactivity. The existing publicly available BWR axial burnup distribution information comes chiefly from datasets describing BWR CRCs [16, 17, 18]. The cycles modeled in these CRCs ended in 1995 and 1996. Consequently, the axial burnup data that can be extracted from the CRCs does not reflect the impact of modern lattices and use of part-length fuel rods. More modern axial burnup distribution data is needed to support accurate evaluation of the impact of axial burnup distributions.

As was mentioned above, additional BWR RCA data are needed to support validation of BWR burned fuel composition calculations. From Reference 30, the 32 BWR RCA samples came from 14 fuel rods that were in 4 assemblies that came from three nuclear plants. The samples came from an 8 x 8 lattice with two small water rods, a 7 x 7 lattice without water rods, and a 6 x 6 lattice without water rods. Fresh fuel initial enrichments varied from 2.53 to 3.91 wt % ^{235}U, had sample burnups ranging from 14.4 to 44.0 GWd/MTU, and had effective water densities ranging from 0.22 to 0.74 g/cm^3. If full range burnup-plus-Gd credit is used, additional BWR RCA data for rods at lower burnups and for rods with gadolinia are needed.

4. RECOMMENDATIONS AND PRIORITIZATION

The primary purpose of this report is to identify work that needs to be performed to support use of BWR burnup credit that credits burnup beyond the peak reactivity point. Section 3 provides discussion of many issues related BWR burnup credit analysis and implementation. Tables 6, 7, and 8 in this section provide a summary of the recommendations and propose priority levels for the recommendations. High priority items are items that should be addressed prior to implementation of full burnup credit. Higher priority items should be resolved before lower priority items are evaluated, so that the impact of the higher priority item resolutions may be considered in resolution of the lower priority items. Medium priority items are items that resolution of which may reduce excessive conservatism associated with some analysis assumptions. Resolution of low priority items is unlikely to have significant effect on the implementation of BWR burnup credit.

Three objectives were identified as "High" priority based on the potential size of their impact on reactivity and on the need to resolve the issues so that the resolution may be included in the study of the lower priority issues. In particular, based on the data presented in Table 20 of NUREG/CR-7157, the modeling of axial burnup distribution will increase k_{eff} by more than 0.09 Δk at assembly average burnups of 60 GWd/MTU compared to using a uniform axial burnup distribution. The impact will likely be greater than 0.1 Δk because the axial burnup profile used in that case is the best-estimate predicted burnup profile and no effort has been made to identify bounding profiles.

The results presented in Figures 10 and 11 and Appendix A indicate that modeling of the moderator density used during depletion can change the resulting k_{eff} value by more than 0.1 Δk when a single moderator density is used for depletion out to 60 GWd/MTU and that the effect may be increased by as much 0.03 Δk if the axial moderator distribution is modeled. As with axial burnup shape modeling, the effect could be significantly larger if since no effort has been made to identify a bounding axial moderator profile.

The results presented in Figure 15 indicate that k_{eff} may be increased by as much as 0.08 Δk if control blades are modeled as fully inserted throughout depletion to 60 GWd/MTU compared to modeling the control blades as fully withdrawn. Due to extensive use of control blades to manage power distribution and core reactivity, a more realistic value is probably less than 0.04 Δk because the rods are typically not fully inserted and usually for only a minor fraction of a cycle. Due to the potential range of the impact on k_{eff}, this area needs to be studied carefully before guidance is provided to applicants.

All medium priority items listed in Table 7 could be addressed by applicants using fairly straightforward analysis techniques. Results from research in these areas could be useful to NRC staff in the review of license applications and may be useful to applicants by providing reference values for comparison and demonstrating techniques useful for analysis supporting license applications.

Table 6. High priority recommendations for BWR burnup credit research

Priority	Research Objective	Activities/tasks
High	Guidance for identification and use of axial burnup distributions	(1) Identify sources of axial burnup distribution data, including data for assemblies with axially dependent features such as blankets and part-length fuel rods (2) Evaluate adequacy of data (3) Obtain or develop additional data if needed (4) Use data to quantify the impact of axial burnup distribution (5) Develop guidance for treatment of axial burnup distributions in BUC
High	Guidance for treatment of axial moderator density distributions	(1) Identify sources of axial moderator density distribution data (2) Evaluate adequacy of data (3) Obtain or develop additional data if needed (4) Use data to quantify the impact of axial moderator density distribution (5) Develop guidance for treatment of axial moderator density distributions
High	Guidance for treatment of control blade usage during depletion	(1) Identify sources of control blade usage data, including control blade designs and control blade operations at power (2) Evaluate adequacy of data (3) Obtain or develop additional data if needed (4) Use data to quantify the impact of control blade usage (5) Develop guidance for treatment of control blade usage
High	Guidance for validation of BWR burned fuel composition calculations	(1) Review adequacy of currently available RCA data to support extension of BWR BUC to higher burnups (2) Update NUREG/CR-7108 work to address BWR fuel with burnups beyond peak reactivity (3) Develop guidance for composition validation of BWR burnup credit beyond peak reactivity
High	Guidance for validation of BWR burnup credit keff calculations	(1) Update NUREG/CR-7109 work to address BWR fuel with burnups beyond peak reactivity (2) Develop guidance for k_{eff} validation of BWR burnup credit beyond peak reactivity

Table 7. Medium priority recommendations for BWR burnup credit research

Medium	Guidance for use of reactor operating parameters in fuel depletion calculations.	(1) Identify reactor operating parameters and ranges. Candidates include fuel temperature, clad temperature, moderator temperature, power density, bypass flow density, etc. (2) Perform sensitivity studies to confirm sensitivity of burned fuel reactivity to reactor operating parameters (3) Develop guidance for selection of reactor operating parameters for fuel depletion calculations
Medium	Guidance for expanded abnormal conditions analysis	Develop BWR-specific guidance for abnormal conditions analysis
Medium	Study and guidance for handling of correlated factors affecting fuel depletion	(1) Identify potentially correlated reactor conditions that may affect used fuel reactivity. Likely candidates are control blade insertion, moderator density, fuel temperatures, and power density. (2) Identify sources of information for correlated reactor conditions (3) Evaluate adequacy of data (4) Obtain or develop additional data if needed (5) Use data to evaluate use of correlated operating data in fuel depletion calculations (6) Develop guidance for use of correlated operating data in fuel depletion
Medium	Guidance for treatment of radial burnup distributions for BWR fuel	(1) Identify sources of radial burnup distribution data (2) Evaluate adequacy of data (3) Obtain or develop additional data if needed (4) Use data to quantify the impact of radial burnup distribution (5) Develop guidance for treatment of radial burnup distributions in BWR burnup credit

Table 8. Low priority recommendations for BWR burnup credit research

Low	Guidance for modeling fuel assembly design variations	(1) Identify sources of information on fuel assembly design variations
		    a) Lattice variations (size, water rods, axial variation of lattice
		    b) Radial and axial enrichment variation
		    c) Gd fuel rod usage (number of rods, position, Gd loading, axial variation)
		    d) Fuel rod dimensions, clad material, pellet density/geometry
		(2) Evaluate adequacy and completeness of data
		(3) Obtain additional data if needed
		(4) Use data to quantify the impact of fuel assembly design variations
		(5) Develop guidance for treatment of fuel assembly design variations and bounding assembly selection

Low	Guidance for treatment of radial burnup distributions for BWR fuel	(1) Identify sources of radial burnup distribution data (2) Evaluate adequacy of data (3) Obtain or develop additional data if needed (4) Use data to quantify the impact of radial burnup distribution (5) Develop guidance for treatment of radial burnup distributions in BWR burnup credit
Low	Guidance for treatment of reactor operating history	(1) Perform study to evaluate the impact of reactor operating history on burned fuel reactivity (2) Develop guidance for treatment of reactor operating history in fuel depletion calculations
Low	Guidance for treatment of fuel assembly changes during irradiation	(1) Identify sources of information on fuel assembly changes that occur during use (2) Evaluate adequacy of data (3) Obtain or develop additional data if needed (4) Evaluate impact of fuel assembly changes on k_{eff} in casks and SFP (5) Develop guidance for treatment of fuel assembly changes during use
Low	Guidance for fuel assembly burnup assignment	(1) Obtain information concerning how BWR assembly burnup values are determined (2) Develop guidance for BWR assembly burnup assignment

5. SUMMARY AND CONCLUSIONS

The work presented in this report is part of a multi-task project sponsored by the NRC that focused on providing support for expanded use of burnup credit for storage and transport of BWR spent nuclear fuel. The first task involved a literature review of prior related work. The literature review was documented in a letter report provided to NRC staff. The second task focused on documenting a reference model, referred to as the GBC-68, that could be used as a basis for comparison in sensitivity studies. Additionally, estimates were produced for the reactivity worth of three minor actinides and 16 fission products as a function of fuel assembly average burnup and post-irradiation cooling time. The reference model and the actinide and fission product worths were documented in NUREG/CR-7157 [15]. The objective of the third task was to identify and prioritize issues associated with BWR burnup credit. This was accomplished through consideration of previously published work and through the execution of some sensitivity studies that were used to quantify the potential impact of an issue on storage or transport system k_{eff} values. The third task resulted in the production of this NUREG/CR report. The final task was to perform a comparison of the currently used peak reactivity method, which credits burnup only to the point at which reactivity peaks as gadolinium is depleted in the BWR fuel assembly, to a simple burnup credit analysis method that provides credit for burnup beyond the peak reactivity point. This comparison was documented in a letter report provided to NRC staff.

The BWR burnup credit issues examined in this report were identified based on knowledge gained from prior BWR and PWR burnup credit studies, experience with real-world implementation of PWR burnup credit, and experience with analyses based on the peak reactivity analysis methods frequently applied to storage of BWR spent fuel. For example, it is clear from experience with PWR burnup credit that at some point as assemblies accumulate burnup it will be necessary to model the axial burnup distribution. This fact was confirmed by performance of one example calculation documented in Section 3.3.3 of NUREG/CR-7157, which showed that beyond an assembly average burnup value somewhere between 10 and 20 GWd/MTU it was non-conservative to ignore the axial burnup distribution in BWR burnup credit analyses.

Numerous computer calculations using the GBC-68 reference model were performed for and documented in this report. In-reactor, two-dimensional fuel pin lattice depletion calculations were performed using the SCALE 6.1 TRITON and STARBUCS programs to generate burned fuel compositions that varied with depletion conditions. The burned fuel compositions were used in CSAS5/KENO-V.a calculations to calculate the k_{eff} for the GBC-68 cask model loaded with spent fuel. The calculated k_{eff} values were used in sensitivity studies to quantify the potential impact of some issues. The sensitivity studies were used as part of the bases for setting priorities for future research needs.

As is discussed in Section 4, the highest priority for future BWR burnup credit research was assigned to modeling of axial fuel burnup distributions in k_{eff} calculations, moderator density during depletion calculations, and control blade usage during depletion calculations. Lack of work in these three areas has the potential to significantly limit implementation of BWR burnup credit beyond the peak reactivity burnup. The results from research and development efforts in these areas are also needed to support study of the other lower priority items listed in Section 4.

It is anticipated that the work documented in this report will be used to design one or more future projects that will focus on providing information and analysis guidance that will be used by applicants in conducting BWR burnup credit analyses and by NRC staff in reviewing BWR burnup credit analyses for spent fuel transportation and storage. Early projects covering axial burnup distribution, reactor moderator density, and control blade usage would produce results that could be included in subsequent work addressing the remainder of the issues. Note that similar work performed for the NRC for PWR burnup credit was conducted over a period of several years. The final product will likely not be a single recommended BWR burnup credit approach, but instead a series of reports that provide guidance for dealing with the relevant issues.

As is noted in Section 3.1, there are several valid approaches to BWR burnup credit implementation. The complexity and level of effort associated with each will be balanced against the applicant's needs. For example, extension of BWR burnup credit to high burnups may be needed to offset the significant loss of margin associated with the continuing degradation of some neutron absorbing panels. In this case a traditional burnup credit loading curve that also credits residual integral gadolinium may be needed. On the other end of the spectrum, it may be possible to demonstrate that, for a particular cask design needing a relatively minor amount of burnup credit, storage of any assemblies with burnups exceeding some low burnup threshold may be acceptable, without regard to its reactor operating history or gadolinium content. It is recommended that future BWR burnup credit research projects be designed to maximize the applicability of the study results to the realistic spectrum of BWR burnup credit applications that are likely to be generated.

6. REFERENCES

1. Spent Fuel Project Office Interim Staff Guidance - 8, Revision 0, "Partial Burnup Credit," U.S. Nuclear Regulatory Commission, May 16, 1999.

2. Spent Fuel Project Office Interim Staff Guidance – 8. Rev. 1, "Limited Burnup Credit," U.S. Nuclear Regulatory Commission, July 30, 1999.

3. Spent Fuel Project Office Interim Staff Guidance – 8, Rev. 2, "Burnup Credit in the Criticality Safety Analyses of PWR Spent Fuel in Transport and Storage Casks," U. S. Nuclear Regulatory Commission, September 27, 2002.

4. Division of Spent Fuel Storage and Transportation Interim Staff Guidance – Rev. 3, "Burnup Credit in the Criticality Safety Analyses of PWR Spent Fuel in Transportation and Storage Casks," U. S. Nuclear Regulatory Commission, September 26, 2012.

5. "Standard Review Plan for Transportation Packages for Spent Nuclear Fuel," NUREG-1617. Washington, D.C.: Nuclear Regulatory Commission, March 2000.

6. B. L. Broadhead, *K-Infinite Trends with Burnup, Enrichment, and Cooling Time for BWR Fuel Assemblies*, ORNL/M-6155, Oak Ridge National Laboratory, August 1999.

7. J. C. Wagner, M. D. DeHart, and B. L. Broadhead, *Investigation of Burnup Credit Modeling Issues Associated with BWR*, ORNL/TM-1999/193, Oak Ridge National Laboratory, October 2000.

8. C. V. Parks, M. D. DeHart, and J. C. Wagner, *Review and Prioritization of Technical Issues Related to Burnup Credit for LWR Fuel*, NUREG/CR-6665 (ORNL/TM-1999/303), U. S. Nuclear Regulatory Committee, Oak Ridge National Laboratory, February 2000.

9. K. W. Cummings and S. E. Turner, "Design of Wet Storage Racks for Spent BWR Fuel," 35608.pdf in *Proc. 2001 ANS Embedded Topical Meeting on Practical Implementation of Nuclear Criticality Safety*, Reno, NV, November 11–15, 2001.

10. C. Casado, J. Sabater, and J. F. Serrano, "Peak Reactivity Characterization and Isotopic Inventory Calculations for BWR Criticality Applications," presented at the *IAEA International Workshop on Advances in Applications of Burnup Credit for Spent Fuel Storage, Transport, Reprocessing, and Disposition*, Córdoba, Spain, October 27-30, 2009.

11. J. Huffer and J. M. Scaglione, *Calculation of Isotopic Bias and Uncertainty for BWR SNF.* CAL-DSU-NU-000003 REV 00A, Office of Civilian Radioactive Waste Management, October 2003.

12. J. M. Scaglione, *44-BWR Waste Package Loading Curve Evaluation*, CAL-DSU-NU-000008 REV 00A, Office of Civilian Radioactive Waste Management, August 2004.

13. J. Huffer, *BWR Axial Profile*, CAL-DSU-NU-000005, REV. 00A, Office of Civilian Radioactive Waste Management, September 2004.

14. *Yucca Mountain Repository License Application,* DOE/RW-0573 Rev. 0, Office of Civilian Radioactive Waste Management, 2008.

15. D. E. Mueller and J. M. Scaglione, Computational Benchmark for Estimation of Reactivity Margin from Fission Products and Minor Actinides in BWR Burnup Credit, NUREG/CR-7157 (ORNL/TM-2012/96), U. S. Nuclear Regulatory Commission, Oak Ridge National Laboratory, June 2012.

16. M. K. Punatar, *Summary Report of Commercial Reactor Criticality Data for Grand Gulf Unit 1*, TDR-UDC-NU-000002, Rev. 00C, Bechtel SAIC Company, LLC, September 2001.

17. D. P. Henderson, *Summary Report of Commercial Reactor Criticality Data for LaSalle Unit 1*, B00000000-01717-5705-00138 REV 00, Civilian Radioactive Waste Management System M&O Contractor, September 1999.

18. D. P. Henderson, *Summary Report of Commercial Reactor Criticality Data for Quad Cities Unit 2*, B00000000-01717-5705-00096, Rev. 01, Civilian Radioactive Waste Management System M&O Contractor, September 1999.

19. M. L. Fensin, *Optimum Boiling Water Reactor Fuel Design Strategies to Enhance Reactor Shutdown by the Standby Liquid Control System*, M.S. Thesis, University of Florida, 2004.

20. J. C. Wagner, M. D. DeHart, and C. V. Parks, *Recommendations for Addressing Axial Burnup in PWR Burnup Credit Analyses,* NUREG/CR-6801 (ORNL/TM-2001/273), U. S. Nuclear Regulatory Commission, Oak Ridge National Laboratory, March 2003.

21. *SCALE: A Comprehensive Modeling and Simulation Suite for Nuclear Safety Analysis and Design*, ORNL/TM-2005/39, Version 6.1, Oak Ridge National Laboratory, Oak Ridge, Tennessee, June 2011. Available from Radiation Safety Information Computational Center at Oak Ridge National Laboratory as CCC-785.

22. J. C. Wagner, *Computational Benchmark for Estimation of Reactivity Margin from Fission Products and Minor Actinides in PWR Burnup Credit*, NUREG/CR-6747 (ORNL/TM-2000/306), U. S. Nuclear Regulatory Commission, Oak Ridge National Laboratory, October 2001.

23. M. D. DeHart, *Parametric Analysis of PWR Spent Fuel Depletion Parameters for Long-Term-Disposal Criticality Safety*, ORNL/TM-1999/99, Oak Ridge National Laboratory, August 1999.

24. M. D. DeHart, *Sensitivity and Parametric Evaluations of Significant Aspects of Burnup Credit for PWR Spent Fuel Packages*, ORNL/TM-12973, Oak Ridge National Laboratory, May 1996.

25. R. J. Cacciapouti and S. Van Volkinburg, "Axial Burnup Profile Database for Pressurized Water Reactors," YAEC-1937, Yankee Atomic Electric Company (May 1997). Available from the Radiation Safety Information Computational Center at Oak Ridge National Laboratory as DLC-201.

26. T. A. Parish and C. H. Chen, *Bounding Axial Profile Analysis for the Topical Report Database*, Nuclear Engineering Dept., Texas A&M University, May 1997.

27. B. B. Bevard, J. C. Wagner, C. V. Parks, and M. Aissa, *Review of Information for Spent Nuclear Fuel Burnup Confirmation*, NUREG/CR-6998, U. S. Nuclear Regulatory Commission, Oak Ridge National Laboratory, December 2009.

28. I. C. Gauld, *Strategies for Application of Isotopic Uncertainties in Burnup Credit*, NUREG/CR-6811 (ORNL/TM-2001/257), U. S. Nuclear Regulatory Commission, Oak Ridge National Laboratory, June 2003.

29. G. Radulescu, I. C. Gauld, G. Ilas, and J. C. Wagner, *An Approach for Validating Actinide and Fission Product Burnup Credit Criticality Safety Analyses – Isotopic Composition Predictions*, NUREG/CR-7108 (ORNL/TM-2011/509),U. S. Nuclear Regulatory Commission, Oak Ridge National Laboratory, April 2012.

30. U. Mertyurek, M. W. Francis, and I. C. Gauld, *SCALE 5 Analysis of BWR Spent Nuclear Fuel Isotopic Compositions for Safety Studies*, ORNL/TM-2010/286, Oak Ridge National Laboratory, December 2010.

31. J. C. Dean and R. W. Tayloe, Jr., *Guide for Validation of Nuclear Criticality Safety Calculational Methodology*, NUREG/CR-6698, U. S. Nuclear Regulatory Commission, Science Applications International Corporation, January 2001.

32. J. M. Scaglione, D. E. Mueller, J. C. Wagner, and W. J. Marshall, *An Approach for Validating Actinide and Fission Product Burnup Credit Criticality Safety Analyses – Criticality (k_{eff}) Predictions*, NUREG/CR-7109 (ORNL/TM-2011/514), U. S. Nuclear Regulatory Commission, Oak Ridge National Laboratory, April 2012.

APPENDIX A: IMPACT OF MODERATOR DENSITY ON FUEL DEPLETION CALCULATIONS

APPENDIX A. IMPACT OF MODERATOR DENSITY ON FUEL DEPLETION CALCULATIONS

Tables and figures are presented below that show how k_{eff} changes with variation of the in-channel moderator density used during the fuel depletion calculations. Results are provided for initial enrichments varying from 2 to 5 wt % ^{235}U, for assembly average burnups ranging up to 60 GWd/MTU, for post-irradiation cooling times of 5, 10, 20 and 40 years, and for primary-actinide-only and actinides plus 16-fission product burnup credit.

Table 9. Sensitivity of k_{eff} to reactor water density – 2 wt %

Burnup (GWd/MTU)	k_{eff} Versus Burnup (GWd/MTU) and Depletion Water Density (g/cm^3)							
	2 wt % ^{235}U, Actinide-Only Burnup Credit, 5-Year Cooling Time							
	0.1	0.2	0.3	0.4	0.5	0.6	0.7	0.8
10	0.7671	0.7602	0.7533	0.7465	0.7401	0.7341	0.7288	0.7237
20	0.7468	0.7348	0.7215	0.7082	0.6955	0.6830	0.6719	0.6614
30	0.7305	0.7147	0.6968	0.6780	0.6598	0.6422	0.6261	0.6108
40	0.7194	0.7010	0.6796	0.6576	0.6359	0.6153	0.5964	0.5789
50	0.7130	0.6929	0.6699	0.6458	0.6225	0.6005	0.5803	0.5624
60	0.7093	0.6896	0.6648	0.6403	0.6162	0.5937	0.5738	0.5561
	2 wt % ^{235}U, Actinide & Fission Product Burnup Credit, 5-Year Cooling Time							
	0.1	0.2	0.3	0.4	0.5	0.6	0.7	0.8
10	0.7330	0.7269	0.7202	0.7139	0.7082	0.7024	0.6972	0.6923
20	0.6955	0.6846	0.6716	0.6592	0.6468	0.6350	0.6246	0.6142
30	0.6663	0.6512	0.6336	0.6160	0.5988	0.5819	0.5667	0.5523
40	0.6447	0.6271	0.6066	0.5860	0.5655	0.5460	0.5286	0.5127
50	0.6300	0.6107	0.5887	0.5663	0.5448	0.5241	0.5059	0.4899
60	0.6195	0.6004	0.5771	0.5540	0.5322	0.5115	0.4935	0.4779
	2 wt % ^{235}U, Actinide-Only Burnup Credit, 10-Year Cooling Time							
	0.1	0.2	0.3	0.4	0.5	0.6	0.7	0.8
10	0.7626	0.7560	0.7490	0.7424	0.7363	0.7303	0.7251	0.7203
20	0.7369	0.7249	0.7112	0.6983	0.6856	0.6734	0.6622	0.6518
30	0.7162	0.7002	0.6818	0.6632	0.6448	0.6274	0.6114	0.5963
40	0.7017	0.6830	0.6613	0.6394	0.6180	0.5974	0.5786	0.5614
50	0.6927	0.6723	0.6490	0.6251	0.6020	0.5802	0.5605	0.5430
60	0.6872	0.6670	0.6425	0.6180	0.5942	0.5723	0.5528	0.5354
	2 wt % ^{235}U, Actinide & Fission Product Burnup Credit, 10-Year Cooling Time							
	0.1	0.2	0.3	0.4	0.5	0.6	0.7	0.8
10	0.7280	0.7216	0.7151	0.7092	0.7035	0.6980	0.6929	0.6884
20	0.6842	0.6729	0.6601	0.6476	0.6356	0.6239	0.6132	0.6033
30	0.6497	0.6342	0.6169	0.5990	0.5819	0.5654	0.5503	0.5362
40	0.6245	0.6065	0.5860	0.5653	0.5450	0.5258	0.5086	0.4932
50	0.6071	0.5877	0.5653	0.5430	0.5217	0.5013	0.4835	0.4680
60	0.5950	0.5755	0.5522	0.5293	0.5078	0.4877	0.4700	0.4548

Table 9. Sensitivity of k_{eff} to reactor water density – 2 wt % (continued)

Burnup (GWd/MTU)	k_{eff} Versus Burnup (GWd/MTU) and Depletion Water Density (g/cm³)							
	2 wt % ²³⁵U, Actinide-Only Burnup Credit, 20-Year Cooling Time							
	0.1	0.2	0.3	0.4	0.5	0.6	0.7	0.8
10	0.7565	0.7500	0.7432	0.7368	0.7309	0.7251	0.7202	0.7155
20	0.7231	0.7107	0.6973	0.6843	0.6714	0.6593	0.6486	0.6386
30	0.6963	0.6794	0.6609	0.6423	0.6239	0.6068	0.5909	0.5762
40	0.6769	0.6574	0.6357	0.6137	0.5922	0.5720	0.5533	0.5365
50	0.6644	0.6435	0.6197	0.5960	0.5732	0.5516	0.5324	0.5153
60	0.6563	0.6354	0.6108	0.5865	0.5634	0.5417	0.5229	0.5060
	2 wt % ²³⁵U, Actinide & Fission Product Burnup Credit, 20-Year Cooling Time							
	0.1	0.2	0.3	0.4	0.5	0.6	0.7	0.8
10	0.7216	0.7156	0.7091	0.7032	0.6976	0.6925	0.6874	0.6829
20	0.6699	0.6583	0.6454	0.6329	0.6209	0.6094	0.5991	0.5895
30	0.6291	0.6131	0.5955	0.5778	0.5608	0.5443	0.5294	0.5156
40	0.5992	0.5807	0.5598	0.5393	0.5192	0.5005	0.4834	0.4681
50	0.5784	0.5585	0.5363	0.5142	0.4930	0.4736	0.4560	0.4410
60	0.5641	0.5442	0.5209	0.4986	0.4772	0.4579	0.4412	0.4265
	2 wt % ²³⁵U, Actinide-Only Burnup Credit, 40-Year Cooling Time							
	0.1	0.2	0.3	0.4	0.5	0.6	0.7	0.8
10	0.7509	0.7444	0.7377	0.7316	0.7258	0.7203	0.7154	0.7107
20	0.7094	0.6968	0.6834	0.6702	0.6578	0.6458	0.6353	0.6252
30	0.6766	0.6592	0.6404	0.6216	0.6038	0.5864	0.5708	0.5560
40	0.6526	0.6322	0.6101	0.5880	0.5667	0.5466	0.5284	0.5119
50	0.6366	0.6145	0.5905	0.5672	0.5443	0.5232	0.5044	0.4878
60	0.6256	0.6041	0.5792	0.5551	0.5320	0.5112	0.4929	0.4767
	2 wt % ²³⁵U, Actinide & Fission Product Burnup Credit, 40-Year Cooling Time							
	0.1	0.2	0.3	0.4	0.5	0.6	0.7	0.8
10	0.7156	0.7099	0.7037	0.6979	0.6924	0.6875	0.6828	0.6782
20	0.6566	0.6449	0.6320	0.6194	0.6076	0.5967	0.5863	0.5768
30	0.6105	0.5940	0.5759	0.5583	0.5411	0.5248	0.5102	0.4966
40	0.5763	0.5571	0.5361	0.5155	0.4957	0.4773	0.4605	0.4455
50	0.5526	0.5321	0.5097	0.4879	0.4668	0.4477	0.4310	0.4166
60	0.5360	0.5157	0.4925	0.4703	0.4496	0.4312	0.4149	0.4009

Table 10. Sensitivity of k_{eff} to reactor water density – 3 wt %

Burnup (GWd/MTU)	k_{eff} Versus Burnup (GWd/MTU) and Depletion Water Density (g/cm^3)							
	3 wt % ^{235}U, Actinide-Only Burnup Credit, 5-Year Cooling Time							
	0.1	0.2	0.3	0.4	0.5	0.6	0.7	0.8
10	0.8528	0.8489	0.8445	0.8408	0.8370	0.8333	0.8303	0.8273
20	0.8212	0.8134	0.8047	0.7957	0.7872	0.7792	0.7719	0.7648
30	0.7922	0.7808	0.7675	0.7534	0.7395	0.7258	0.7131	0.7010
40	0.7683	0.7534	0.7354	0.7170	0.6982	0.6794	0.6620	0.6453
50	0.7499	0.7321	0.7110	0.6885	0.6659	0.6437	0.6231	0.6038
60	0.7363	0.7176	0.6941	0.6695	0.6447	0.6207	0.5989	0.5790
	3 wt % ^{235}U, Actinide & Fission Product Burnup Credit, 5-Year Cooling Time							
	0.1	0.2	0.3	0.4	0.5	0.6	0.7	0.8
10	0.8179	0.8145	0.8110	0.8073	0.8038	0.8009	0.7979	0.7953
20	0.7690	0.7619	0.7534	0.7453	0.7370	0.7297	0.7225	0.7155
30	0.7263	0.7152	0.7021	0.6889	0.6754	0.6625	0.6503	0.6390
40	0.6911	0.6767	0.6595	0.6415	0.6232	0.6057	0.5892	0.5734
50	0.6641	0.6467	0.6262	0.6049	0.5837	0.5628	0.5435	0.5259
60	0.6435	0.6251	0.6029	0.5794	0.5562	0.5343	0.5144	0.4964
	3 wt % ^{235}U, Actinide-Only Burnup Credit, 10-Year Cooling Time							
	0.1	0.2	0.3	0.4	0.5	0.6	0.7	0.8
10	0.8501	0.8465	0.8420	0.8385	0.8348	0.8314	0.8285	0.8255
20	0.8139	0.8060	0.7976	0.7889	0.7803	0.7723	0.7651	0.7584
30	0.7810	0.7691	0.7556	0.7416	0.7278	0.7147	0.7022	0.6902
40	0.7535	0.7381	0.7202	0.7014	0.6826	0.6640	0.6468	0.6302
50	0.7320	0.7137	0.6924	0.6701	0.6475	0.6253	0.6048	0.5860
60	0.7162	0.6968	0.6735	0.6486	0.6240	0.6004	0.5790	0.5594
	3 wt % ^{235}U, Actinide & Fission Product Burnup Credit, 10-Year Cooling Time							
	0.1	0.2	0.3	0.4	0.5	0.6	0.7	0.8
10	0.8149	0.8113	0.8078	0.8042	0.8012	0.7982	0.7951	0.7928
20	0.7607	0.7532	0.7452	0.7368	0.7290	0.7213	0.7145	0.7081
30	0.7131	0.7016	0.6886	0.6751	0.6620	0.6489	0.6369	0.6258
40	0.6736	0.6588	0.6414	0.6235	0.6055	0.5881	0.5714	0.5560
50	0.6432	0.6257	0.6048	0.5835	0.5623	0.5418	0.5228	0.5052
60	0.6203	0.6019	0.5791	0.5557	0.5325	0.5111	0.4915	0.4742

Table 10. Sensitivity of k_{eff} to reactor water density – 3 wt % (continued)

Burnup (GWd/MTU)	k_{eff} Versus Burnup (GWd/MTU) and Depletion Water Density (g/cm^3)							
	3 wt % ^{235}U, Actinide-Only Burnup Credit, 20-Year Cooling Time							
	0.1	0.2	0.3	0.4	0.5	0.6	0.7	0.8
10	0.8466	0.8426	0.8389	0.8351	0.8319	0.8286	0.8256	0.8230
20	0.8040	0.7963	0.7874	0.7792	0.7709	0.7631	0.7561	0.7496
30	0.7652	0.7531	0.7393	0.7255	0.7117	0.6986	0.6863	0.6746
40	0.7323	0.7164	0.6984	0.6793	0.6608	0.6424	0.6253	0.6091
50	0.7066	0.6880	0.6663	0.6437	0.6213	0.5994	0.5791	0.5604
60	0.6877	0.6676	0.6438	0.6191	0.5947	0.5715	0.5502	0.5314
	3 wt % ^{235}U, Actinide & Fission Product Burnup Credit, 20-Year Cooling Time							
	0.1	0.2	0.3	0.4	0.5	0.6	0.7	0.8
10	0.8108	0.8076	0.8042	0.8007	0.7978	0.7947	0.7920	0.7897
20	0.7502	0.7427	0.7346	0.7265	0.7188	0.7114	0.7045	0.6984
30	0.6963	0.6846	0.6717	0.6584	0.6451	0.6326	0.6206	0.6095
40	0.6520	0.6365	0.6190	0.6010	0.5834	0.5657	0.5495	0.5342
50	0.6176	0.5993	0.5784	0.5572	0.5363	0.5157	0.4968	0.4799
60	0.5915	0.5725	0.5496	0.5266	0.5038	0.4829	0.4638	0.4467
	3 wt % ^{235}U, Actinide-Only Burnup Credit, 40-Year Cooling Time							
	0.1	0.2	0.3	0.4	0.5	0.6	0.7	0.8
10	0.8431	0.8395	0.8358	0.8321	0.8288	0.8257	0.8232	0.8204
20	0.7945	0.7863	0.7779	0.7696	0.7615	0.7541	0.7473	0.7410
30	0.7496	0.7374	0.7236	0.7097	0.6960	0.6830	0.6709	0.6595
40	0.7117	0.6955	0.6767	0.6581	0.6393	0.6214	0.6044	0.5881
50	0.6819	0.6626	0.6401	0.6177	0.5954	0.5737	0.5539	0.5355
60	0.6597	0.6388	0.6147	0.5895	0.5654	0.5426	0.5220	0.5034
	3 wt % ^{235}U, Actinide & Fission Product Burnup Credit, 40-Year Cooling Time							
	0.1	0.2	0.3	0.4	0.5	0.6	0.7	0.8
10	0.8074	0.8042	0.8009	0.7978	0.7949	0.7921	0.7896	0.7872
20	0.7407	0.7333	0.7252	0.7173	0.7097	0.7027	0.6958	0.6899
30	0.6813	0.6695	0.6564	0.6428	0.6299	0.6175	0.6057	0.5948
40	0.6325	0.6167	0.5988	0.5808	0.5630	0.5456	0.5299	0.5148
50	0.5945	0.5758	0.5545	0.5330	0.5122	0.4920	0.4740	0.4572
60	0.5658	0.5461	0.5232	0.5000	0.4776	0.4570	0.4386	0.4223

Table 11. Sensitivity of k_{eff} to reactor water density – 4 wt %

Burnup (GWd/MTU)	k_{eff} Versus Burnup (GWd/MTU) and Depletion Water Density (g/cm³)							
	4 wt % ²³⁵U, Actinide-Only Burnup Credit, 5-Year Cooling Time							
	0.1	0.2	0.3	0.4	0.5	0.6	0.7	0.8
10	0.9158	0.9134	0.9107	0.9082	0.9062	0.9039	0.9020	0.9001
20	0.8823	0.8770	0.8714	0.8659	0.8603	0.8552	0.8506	0.8461
30	0.8498	0.8418	0.8328	0.8232	0.8139	0.8045	0.7958	0.7876
40	0.8200	0.8093	0.7958	0.7818	0.7677	0.7536	0.7404	0.7275
50	0.7941	0.7802	0.7632	0.7445	0.7254	0.7066	0.6885	0.6708
60	0.7730	0.7561	0.7359	0.7133	0.6901	0.6678	0.6460	0.6254
	4 wt % ²³⁵U, Actinide & Fission Product Burnup Credit, 5-Year Cooling Time							
	0.1	0.2	0.3	0.4	0.5	0.6	0.7	0.8
10	0.8808	0.8788	0.8769	0.8748	0.8728	0.8713	0.8694	0.8678
20	0.8302	0.8257	0.8202	0.8153	0.8100	0.8053	0.8007	0.7966
30	0.7832	0.7758	0.7672	0.7578	0.7485	0.7397	0.7316	0.7234
40	0.7419	0.7310	0.7181	0.7042	0.6905	0.6771	0.6639	0.6516
50	0.7061	0.6921	0.6755	0.6574	0.6391	0.6208	0.6038	0.5874
60	0.6772	0.6607	0.6406	0.6189	0.5969	0.5759	0.5556	0.5366
	4 wt % ²³⁵U, Actinide-Only Burnup Credit, 10-Year Cooling Time							
	0.1	0.2	0.3	0.4	0.5	0.6	0.7	0.8
10	0.9140	0.9119	0.9092	0.9068	0.9046	0.9027	0.9007	0.8991
20	0.8770	0.8719	0.8662	0.8608	0.8555	0.8506	0.8459	0.8416
30	0.8408	0.8328	0.8236	0.8141	0.8048	0.7958	0.7875	0.7795
40	0.8077	0.7964	0.7832	0.7691	0.7549	0.7409	0.7279	0.7155
50	0.7787	0.7643	0.7468	0.7283	0.7093	0.6906	0.6725	0.6555
60	0.7547	0.7374	0.7171	0.6943	0.6712	0.6488	0.6274	0.6073
	4 wt % ²³⁵U, Actinide & Fission Product Burnup Credit, 10-Year Cooling Time							
	0.1	0.2	0.3	0.4	0.5	0.6	0.7	0.8
10	0.8786	0.8768	0.8747	0.8728	0.8711	0.8692	0.8677	0.8664
20	0.8237	0.8194	0.8141	0.8089	0.8040	0.7996	0.7951	0.7910
30	0.7727	0.7650	0.7561	0.7471	0.7382	0.7295	0.7211	0.7135
40	0.7270	0.7160	0.7029	0.6892	0.6752	0.6619	0.6492	0.6368
50	0.6881	0.6739	0.6567	0.6384	0.6202	0.6024	0.5851	0.5686
60	0.6561	0.6391	0.6191	0.5973	0.5751	0.5542	0.5342	0.5155

Table 11. Sensitivity of k_{eff} to reactor water density – 4 wt % (continued)

Burnup (GWd/MTU)	k_{eff} Versus Burnup (GWd/MTU) and Depletion Water Density (g/cm³)							
	4 wt % ²³⁵U, Actinide-Only Burnup Credit, 20-Year Cooling Time							
	0.1	0.2	0.3	0.4	0.5	0.6	0.7	0.8
10	0.9118	0.9094	0.9068	0.9049	0.9026	0.9009	0.8990	0.8975
20	0.8698	0.8646	0.8592	0.8538	0.8488	0.8440	0.8395	0.8356
30	0.8287	0.8203	0.8111	0.8017	0.7927	0.7838	0.7753	0.7676
40	0.7906	0.7785	0.7652	0.7511	0.7372	0.7236	0.7106	0.6984
50	0.7570	0.7420	0.7242	0.7055	0.6868	0.6682	0.6504	0.6336
60	0.7294	0.7112	0.6904	0.6676	0.6450	0.6225	0.6013	0.5814
	4 wt % ²³⁵U, Actinide & Fission Product Burnup Credit, 20-Year Cooling Time							
	0.1	0.2	0.3	0.4	0.5	0.6	0.7	0.8
10	0.8760	0.8742	0.8722	0.8702	0.8688	0.8673	0.8657	0.8646
20	0.8160	0.8114	0.8064	0.8015	0.7968	0.7923	0.7882	0.7843
30	0.7596	0.7517	0.7429	0.7339	0.7250	0.7165	0.7086	0.7009
40	0.7089	0.6974	0.6842	0.6706	0.6568	0.6437	0.6310	0.6191
50	0.6656	0.6507	0.6331	0.6153	0.5971	0.5794	0.5622	0.5463
60	0.6302	0.6125	0.5922	0.5704	0.5487	0.5279	0.5081	0.4900
	4 wt % ²³⁵U, Actinide-Only Burnup Credit, 40-Year Cooling Time							
	0.1	0.2	0.3	0.4	0.5	0.6	0.7	0.8
10	0.9095	0.9074	0.9050	0.9029	0.9010	0.8993	0.8974	0.8962
20	0.8629	0.8577	0.8522	0.8471	0.8422	0.8376	0.8335	0.8297
30	0.8166	0.8081	0.7990	0.7893	0.7808	0.7722	0.7641	0.7565
40	0.7737	0.7615	0.7477	0.7338	0.7202	0.7066	0.6938	0.6817
50	0.7360	0.7201	0.7020	0.6835	0.6646	0.6463	0.6285	0.6118
60	0.7043	0.6856	0.6644	0.6413	0.6186	0.5965	0.5756	0.5559
	4 wt % ²³⁵U, Actinide & Fission Product Burnup Credit, 40-Year Cooling Time							
	0.1	0.2	0.3	0.4	0.5	0.6	0.7	0.8
10	0.8738	0.8720	0.8703	0.8688	0.8670	0.8655	0.8641	0.8630
20	0.8092	0.8045	0.7997	0.7948	0.7905	0.7863	0.7820	0.7786
30	0.7479	0.7398	0.7308	0.7219	0.7132	0.7050	0.6972	0.6899
40	0.6926	0.6806	0.6674	0.6538	0.6401	0.6273	0.6150	0.6031
50	0.6456	0.6302	0.6124	0.5943	0.5762	0.5585	0.5417	0.5261
60	0.6071	0.5885	0.5682	0.5462	0.5244	0.5043	0.4849	0.4672

Table 12. Sensitivity of k_{eff} to reactor water density – 5 wt %

Burnup (GWd/MTU)	k_{eff} Versus Burnup (GWd/MTU) and Depletion Water Density (g/cm^3)							
	5 wt % ^{235}U, Actinide-Only Burnup Credit, 5-Year Cooling Time							
	0.1	0.2	0.3	0.4	0.5	0.6	0.7	0.8
10	0.9635	0.9620	0.9603	0.9587	0.9570	0.9560	0.9547	0.9535
20	0.9309	0.9277	0.9238	0.9203	0.9165	0.9131	0.9100	0.9072
30	0.8992	0.8937	0.8873	0.8809	0.8743	0.8683	0.8625	0.8571
40	0.8681	0.8604	0.8511	0.8410	0.8308	0.8211	0.8117	0.8026
50	0.8388	0.8283	0.8159	0.8019	0.7873	0.7732	0.7592	0.7458
60	0.8126	0.7997	0.7833	0.7652	0.7460	0.7273	0.7084	0.6907
	5 wt % ^{235}U, Actinide & Fission Product Burnup Credit, 5-Year Cooling Time							
	0.1	0.2	0.3	0.4	0.5	0.6	0.7	0.8
10	0.9286	0.9276	0.9265	0.9254	0.9241	0.9233	0.9224	0.9214
20	0.8792	0.8762	0.8733	0.8699	0.8667	0.8637	0.8609	0.8583
30	0.8332	0.8277	0.8220	0.8156	0.8095	0.8038	0.7983	0.7930
40	0.7896	0.7819	0.7727	0.7632	0.7532	0.7435	0.7343	0.7255
50	0.7499	0.7395	0.7270	0.7131	0.6988	0.6849	0.6713	0.6582
60	0.7151	0.7023	0.6856	0.6679	0.6493	0.6310	0.6130	0.5959
	5 wt % ^{235}U, Actinide-Only Burnup Credit, 10-Year Cooling Time							
	0.1	0.2	0.3	0.4	0.5	0.6	0.7	0.8
10	0.9624	0.9609	0.9593	0.9579	0.9563	0.9550	0.9538	0.9528
20	0.9270	0.9236	0.9199	0.9164	0.9128	0.9098	0.9067	0.9041
30	0.8919	0.8865	0.8803	0.8740	0.8674	0.8614	0.8561	0.8509
40	0.8578	0.8498	0.8404	0.8306	0.8206	0.8111	0.8018	0.7930
50	0.8257	0.8147	0.8021	0.7882	0.7738	0.7596	0.7459	0.7325
60	0.7969	0.7832	0.7664	0.7485	0.7294	0.7107	0.6922	0.6744
	5 wt % ^{235}U, Actinide & Fission Product Burnup Credit, 10-Year Cooling Time							
	0.1	0.2	0.3	0.4	0.5	0.6	0.7	0.8
10	0.9269	0.9260	0.9250	0.9238	0.9230	0.9220	0.9211	0.9202
20	0.8744	0.8716	0.8684	0.8652	0.8623	0.8593	0.8567	0.8541
30	0.8243	0.8189	0.8133	0.8073	0.8013	0.7956	0.7902	0.7851
40	0.7772	0.7695	0.7602	0.7505	0.7408	0.7311	0.7219	0.7136
50	0.7342	0.7234	0.7106	0.6968	0.6826	0.6687	0.6551	0.6422
60	0.6963	0.6829	0.6661	0.6481	0.6298	0.6115	0.5939	0.5766

Table 12. Sensitivity of k_{eff} to reactor water density – 5 wt % (continued)

Burnup (GWd/MTU)	k_{eff} Versus Burnup (GWd/MTU) and Depletion Water Density (g/cm^3)							
	5 wt % ^{235}U, Actinide-Only Burnup Credit, 20-Year Cooling Time							
	0.1	0.2	0.3	0.4	0.5	0.6	0.7	0.8
10	0.9611	0.9595	0.9577	0.9562	0.9549	0.9538	0.9528	0.9517
20	0.9218	0.9182	0.9147	0.9112	0.9080	0.9050	0.9022	0.8998
30	0.8823	0.8765	0.8702	0.8641	0.8579	0.8522	0.8469	0.8418
40	0.8437	0.8353	0.8261	0.8160	0.8061	0.7966	0.7878	0.7794
50	0.8073	0.7958	0.7830	0.7692	0.7549	0.7408	0.7273	0.7146
60	0.7742	0.7604	0.7433	0.7251	0.7062	0.6874	0.6693	0.6516
	5 wt % ^{235}U, Actinide & Fission Product Burnup Credit, 20-Year Cooling Time							
	0.1	0.2	0.3	0.4	0.5	0.6	0.7	0.8
10	0.9252	0.9243	0.9232	0.9225	0.9213	0.9205	0.9198	0.9189
20	0.8684	0.8657	0.8626	0.8597	0.8566	0.8540	0.8515	0.8492
30	0.8139	0.8085	0.8027	0.7967	0.7909	0.7856	0.7804	0.7754
40	0.7621	0.7540	0.7447	0.7352	0.7254	0.7160	0.7074	0.6990
50	0.7148	0.7037	0.6904	0.6769	0.6626	0.6489	0.6355	0.6230
60	0.6733	0.6590	0.6420	0.6242	0.6058	0.5878	0.5702	0.5534
	5 wt % ^{235}U, Actinide-Only Burnup Credit, 40-Year Cooling Time							
	0.1	0.2	0.3	0.4	0.5	0.6	0.7	0.8
10	0.9594	0.9578	0.9564	0.9550	0.9539	0.9526	0.9518	0.9510
20	0.9165	0.9130	0.9096	0.9063	0.9032	0.9003	0.8980	0.8955
30	0.8731	0.8671	0.8606	0.8546	0.8488	0.8433	0.8380	0.8335
40	0.8301	0.8213	0.8115	0.8022	0.7922	0.7832	0.7744	0.7662
50	0.7896	0.7776	0.7644	0.7504	0.7362	0.7224	0.7093	0.6966
60	0.7527	0.7380	0.7204	0.7022	0.6832	0.6648	0.6468	0.6295
	5 wt % ^{235}U, Actinide & Fission Product Burnup Credit, 40-Year Cooling Time							
	0.1	0.2	0.3	0.4	0.5	0.6	0.7	0.8
10	0.9237	0.9228	0.9217	0.9210	0.9201	0.9192	0.9186	0.9180
20	0.8634	0.8605	0.8575	0.8547	0.8517	0.8495	0.8473	0.8449
30	0.8045	0.7993	0.7933	0.7876	0.7819	0.7765	0.7718	0.7675
40	0.7489	0.7403	0.7311	0.7214	0.7118	0.7029	0.6944	0.6862
50	0.6977	0.6861	0.6729	0.6591	0.6450	0.6318	0.6185	0.6063
60	0.6529	0.6382	0.6209	0.6029	0.5845	0.5666	0.5496	0.5331

Table 13. Sensitivity of Δk_{eff} to reactor water density – 2 wt %

Burnup (GWd/MTU)	Δk_{eff} Versus Burnup (GWd/MTU) and Depletion Water Density (g/cm^3)							
	2 wt % ^{235}U, Actinide-Only Burnup Credit, 5-Year Cooling Time							
	0.1	0.2	0.3	0.4	0.5	0.6	0.7	0.8
10	0.0330	0.0262	0.0193	0.0125	0.0060	0	-0.0053	-0.0104
20	0.0637	0.0518	0.0385	0.0252	0.0124	0	-0.0111	-0.0217
30	0.0882	0.0724	0.0545	0.0357	0.0175	0	-0.0162	-0.0315
40	0.1041	0.0857	0.0643	0.0423	0.0206	0	-0.0189	-0.0364
50	0.1125	0.0924	0.0694	0.0453	0.0220	0	-0.0202	-0.0381
60	0.1157	0.0959	0.0712	0.0466	0.0225	0	-0.0199	-0.0376
	2 wt % ^{235}U, Actinide & Fission Product Burnup Credit, 5-Year Cooling Time							
	0.1	0.2	0.3	0.4	0.5	0.6	0.7	0.8
10	0.0306	0.0245	0.0178	0.0115	0.0058	0	-0.0052	-0.0101
20	0.0604	0.0496	0.0366	0.0242	0.0118	0	-0.0104	-0.0208
30	0.0843	0.0692	0.0517	0.0341	0.0168	0	-0.0152	-0.0296
40	0.0987	0.0811	0.0606	0.0400	0.0195	0	-0.0174	-0.0333
50	0.1059	0.0865	0.0645	0.0421	0.0206	0	-0.0183	-0.0343
60	0.1081	0.0889	0.0657	0.0426	0.0207	0	-0.0180	-0.0336
	2 wt % ^{235}U, Actinide-Only Burnup Credit, 10-Year Cooling Time							
	0.1	0.2	0.3	0.4	0.5	0.6	0.7	0.8
10	0.0323	0.0257	0.0187	0.0121	0.0059	0	-0.0053	-0.0100
20	0.0634	0.0515	0.0378	0.0248	0.0122	0	-0.0112	-0.0217
30	0.0888	0.0728	0.0545	0.0358	0.0174	0	-0.0160	-0.0311
40	0.1043	0.0856	0.0639	0.0420	0.0206	0	-0.0187	-0.0360
50	0.1126	0.0922	0.0689	0.0450	0.0219	0	-0.0197	-0.0372
60	0.1149	0.0947	0.0703	0.0457	0.0219	0	-0.0195	-0.0369
	2 wt % ^{235}U, Actinide & Fission Product Burnup Credit, 10-Year Cooling Time							
	0.1	0.2	0.3	0.4	0.5	0.6	0.7	0.8
10	0.0300	0.0236	0.0171	0.0112	0.0055	0	-0.0051	-0.0096
20	0.0604	0.0490	0.0363	0.0238	0.0118	0	-0.0107	-0.0206
30	0.0843	0.0688	0.0515	0.0336	0.0165	0	-0.0151	-0.0292
40	0.0987	0.0807	0.0602	0.0395	0.0192	0	-0.0171	-0.0326
50	0.1058	0.0864	0.0640	0.0417	0.0204	0	-0.0178	-0.0334
60	0.1072	0.0878	0.0645	0.0416	0.0200	0	-0.0177	-0.0330

Table 13. Sensitivity of Δk_{eff} to reactor water density – 2 wt % (continued)

	Δk_{eff} Versus Burnup (GWd/MTU) and Depletion Water Density (g/cm^3)							
Burnup (GWd/MTU)	2 wt % ^{235}U, Actinide-Only Burnup Credit, 20-Year Cooling Time							
	0.1	0.2	0.3	0.4	0.5	0.6	0.7	0.8
10	0.0315	0.0250	0.0182	0.0118	0.0058	0	-0.0049	-0.0095
20	0.0638	0.0514	0.0379	0.0250	0.0121	0	-0.0108	-0.0207
30	0.0895	0.0726	0.0542	0.0356	0.0171	0	-0.0158	-0.0306
40	0.1049	0.0855	0.0638	0.0417	0.0202	0	-0.0186	-0.0355
50	0.1128	0.0919	0.0681	0.0444	0.0216	0	-0.0192	-0.0363
60	0.1147	0.0938	0.0692	0.0448	0.0218	0	-0.0188	-0.0356
	2 wt % ^{235}U, Actinide & Fission Product Burnup Credit, 20-Year Cooling Time							
	0.1	0.2	0.3	0.4	0.5	0.6	0.7	0.8
10	0.0291	0.0232	0.0166	0.0107	0.0051	0	-0.0050	-0.0095
20	0.0605	0.0489	0.0360	0.0235	0.0115	0	-0.0103	-0.0199
30	0.0848	0.0688	0.0512	0.0335	0.0164	0	-0.0149	-0.0288
40	0.0987	0.0802	0.0593	0.0388	0.0187	0	-0.0171	-0.0324
50	0.1048	0.0849	0.0627	0.0406	0.0194	0	-0.0176	-0.0326
60	0.1063	0.0863	0.0630	0.0407	0.0193	0	-0.0167	-0.0314
	2 wt % ^{235}U, Actinide-Only Burnup Credit, 40-Year Cooling Time							
	0.1	0.2	0.3	0.4	0.5	0.6	0.7	0.8
10	0.0306	0.0241	0.0174	0.0114	0.0055	0	-0.0049	-0.0095
20	0.0636	0.0511	0.0376	0.0245	0.0120	0	-0.0105	-0.0205
30	0.0903	0.0728	0.0540	0.0353	0.0174	0	-0.0156	-0.0304
40	0.1059	0.0856	0.0634	0.0414	0.0201	0	-0.0183	-0.0347
50	0.1134	0.0914	0.0674	0.0440	0.0211	0	-0.0188	-0.0353
60	0.1144	0.0929	0.0680	0.0439	0.0208	0	-0.0184	-0.0345
	2 wt % ^{235}U, Actinide & Fission Product Burnup Credit, 40-Year Cooling Time							
	0.1	0.2	0.3	0.4	0.5	0.6	0.7	0.8
10	0.0281	0.0224	0.0162	0.0104	0.0049	0	-0.0047	-0.0092
20	0.0600	0.0482	0.0354	0.0227	0.0110	0	-0.0103	-0.0198
30	0.0856	0.0692	0.0511	0.0335	0.0163	0	-0.0146	-0.0282
40	0.0990	0.0798	0.0588	0.0381	0.0183	0	-0.0168	-0.0319
50	0.1048	0.0844	0.0620	0.0402	0.0191	0	-0.0168	-0.0311
60	0.1048	0.0846	0.0613	0.0391	0.0185	0	-0.0162	-0.0303

Table 14. Sensitivity of Δk_{eff} to reactor water density – 3 wt %

Burnup (GWd/MTU)	Δk_{eff} Versus Burnup (GWd/MTU) and Depletion Water Density (g/cm^3)							
	3 wt % ^{235}U, Actinide-Only Burnup Credit, 5-Year Cooling Time							
	0.1	0.2	0.3	0.4	0.5	0.6	0.7	0.8
10	0.0195	0.0156	0.0112	0.0075	0.0037	0	-0.0031	-0.0060
20	0.0419	0.0342	0.0254	0.0165	0.0080	0	-0.0073	-0.0144
30	0.0664	0.0550	0.0417	0.0276	0.0138	0	-0.0126	-0.0247
40	0.0889	0.0740	0.0560	0.0376	0.0188	0	-0.0175	-0.0342
50	0.1062	0.0884	0.0673	0.0448	0.0222	0	-0.0206	-0.0399
60	0.1156	0.0969	0.0734	0.0488	0.0240	0	-0.0218	-0.0417
	3 wt % ^{235}U, Actinide & Fission Product Burnup Credit, 5-Year Cooling Time							
	0.1	0.2	0.3	0.4	0.5	0.6	0.7	0.8
10	0.0170	0.0136	0.0101	0.0064	0.0029	0	-0.0030	-0.0056
20	0.0394	0.0322	0.0238	0.0157	0.0074	0	-0.0072	-0.0141
30	0.0639	0.0527	0.0396	0.0264	0.0129	0	-0.0122	-0.0235
40	0.0855	0.0710	0.0538	0.0358	0.0176	0	-0.0165	-0.0323
50	0.1013	0.0839	0.0634	0.0421	0.0209	0	-0.0193	-0.0369
60	0.1092	0.0908	0.0686	0.0451	0.0219	0	-0.0199	-0.0379
	3 wt % ^{235}U, Actinide-Only Burnup Credit, 10-Year Cooling Time							
	0.1	0.2	0.3	0.4	0.5	0.6	0.7	0.8
10	0.0186	0.0151	0.0106	0.0071	0.0033	0	-0.0030	-0.0059
20	0.0416	0.0337	0.0253	0.0166	0.0080	0	-0.0072	-0.0139
30	0.0663	0.0544	0.0410	0.0269	0.0131	0	-0.0125	-0.0245
40	0.0895	0.0741	0.0562	0.0374	0.0186	0	-0.0172	-0.0338
50	0.1068	0.0884	0.0672	0.0448	0.0223	0	-0.0205	-0.0392
60	0.1158	0.0964	0.0730	0.0482	0.0236	0	-0.0214	-0.0410
	3 wt % ^{235}U, Actinide & Fission Product Burnup Credit, 10-Year Cooling Time							
	0.1	0.2	0.3	0.4	0.5	0.6	0.7	0.8
10	0.0167	0.0131	0.0096	0.0060	0.0030	0	-0.0031	-0.0054
20	0.0394	0.0319	0.0238	0.0155	0.0077	0	-0.0068	-0.0133
30	0.0642	0.0528	0.0398	0.0263	0.0132	0	-0.0119	-0.0231
40	0.0855	0.0707	0.0533	0.0354	0.0174	0	-0.0167	-0.0321
50	0.1015	0.0839	0.0630	0.0417	0.0205	0	-0.0190	-0.0366
60	0.1092	0.0907	0.0680	0.0445	0.0214	0	-0.0196	-0.0370

Table 14. Sensitivity of Δk_{eff} to reactor water density – 3 wt % (continued)

Burnup (GWd/MTU)	Δk_{eff} Versus Burnup (GWd/MTU) and Depletion Water Density (g/cm^3)							
	3 wt % ^{235}U, Actinide-Only Burnup Credit, 20-Year Cooling Time							
	0.1	0.2	0.3	0.4	0.5	0.6	0.7	0.8
10	0.0181	0.0140	0.0103	0.0066	0.0033	0	-0.0030	-0.0055
20	0.0410	0.0333	0.0244	0.0161	0.0078	0	-0.0070	-0.0134
30	0.0666	0.0545	0.0408	0.0269	0.0132	0	-0.0123	-0.0239
40	0.0899	0.0740	0.0560	0.0369	0.0184	0	-0.0171	-0.0333
50	0.1072	0.0886	0.0669	0.0443	0.0219	0	-0.0203	-0.0390
60	0.1162	0.0961	0.0723	0.0476	0.0232	0	-0.0213	-0.0401
	3 wt % ^{235}U, Actinide & Fission Product Burnup Credit, 20-Year Cooling Time							
	0.1	0.2	0.3	0.4	0.5	0.6	0.7	0.8
10	0.0161	0.0129	0.0095	0.0059	0.0031	0	-0.0027	-0.0050
20	0.0387	0.0313	0.0232	0.0150	0.0074	0	-0.0069	-0.0130
30	0.0638	0.0521	0.0391	0.0258	0.0125	0	-0.0120	-0.0230
40	0.0863	0.0708	0.0533	0.0353	0.0177	0	-0.0162	-0.0315
50	0.1019	0.0836	0.0627	0.0415	0.0206	0	-0.0189	-0.0358
60	0.1087	0.0896	0.0668	0.0438	0.0209	0	-0.0191	-0.0362
	3 wt % ^{235}U, Actinide-Only Burnup Credit, 40-Year Cooling Time							
	0.1	0.2	0.3	0.4	0.5	0.6	0.7	0.8
10	0.0175	0.0138	0.0101	0.0064	0.0032	0	-0.0025	-0.0053
20	0.0404	0.0323	0.0239	0.0156	0.0074	0	-0.0067	-0.0131
30	0.0666	0.0544	0.0406	0.0267	0.0130	0	-0.0121	-0.0236
40	0.0903	0.0741	0.0554	0.0367	0.0179	0	-0.0170	-0.0332
50	0.1081	0.0889	0.0664	0.0440	0.0217	0	-0.0198	-0.0382
60	0.1171	0.0962	0.0721	0.0469	0.0228	0	-0.0206	-0.0392
	3 wt % ^{235}U, Actinide & Fission Product Burnup Credit, 40-Year Cooling Time							
	0.1	0.2	0.3	0.4	0.5	0.6	0.7	0.8
10	0.0153	0.0122	0.0088	0.0057	0.0028	0	-0.0025	-0.0049
20	0.0381	0.0306	0.0225	0.0147	0.0070	0	-0.0069	-0.0128
30	0.0639	0.0520	0.0389	0.0253	0.0124	0	-0.0117	-0.0227
40	0.0870	0.0711	0.0532	0.0352	0.0174	0	-0.0157	-0.0307
50	0.1025	0.0838	0.0625	0.0410	0.0202	0	-0.0180	-0.0348
60	0.1088	0.0891	0.0662	0.0430	0.0206	0	-0.0185	-0.0347

Table 15. Sensitivity of Δk_{eff} to reactor water density – 4 wt %

Burnup (GWd/MTU)	Δk_{eff} Versus Burnup (GWd/MTU) and Depletion Water Density (g/cm³)							
	4 wt % ²³⁵U, Actinide-Only Burnup Credit, 5-Year Cooling Time							
	0.1	0.2	0.3	0.4	0.5	0.6	0.7	0.8
10	0.0118	0.0095	0.0068	0.0043	0.0022	0	-0.0020	-0.0039
20	0.0272	0.0219	0.0162	0.0107	0.0051	0	-0.0046	-0.0090
30	0.0453	0.0373	0.0284	0.0187	0.0094	0	-0.0087	-0.0169
40	0.0664	0.0557	0.0422	0.0282	0.0141	0	-0.0132	-0.0261
50	0.0875	0.0736	0.0566	0.0379	0.0189	0	-0.0181	-0.0358
60	0.1052	0.0882	0.0681	0.0455	0.0222	0	-0.0218	-0.0424
	4 wt % ²³⁵U, Actinide & Fission Product Burnup Credit, 5-Year Cooling Time							
	0.1	0.2	0.3	0.4	0.5	0.6	0.7	0.8
10	0.0095	0.0076	0.0056	0.0035	0.0016	0	-0.0019	-0.0035
20	0.0249	0.0204	0.0149	0.0100	0.0047	0	-0.0046	-0.0086
30	0.0436	0.0361	0.0276	0.0181	0.0089	0	-0.0081	-0.0163
40	0.0648	0.0539	0.0411	0.0272	0.0134	0	-0.0131	-0.0255
50	0.0853	0.0712	0.0547	0.0366	0.0183	0	-0.0171	-0.0334
60	0.1013	0.0848	0.0647	0.0430	0.0210	0	-0.0203	-0.0393
	4 wt % ²³⁵U, Actinide-Only Burnup Credit, 10-Year Cooling Time							
	0.1	0.2	0.3	0.4	0.5	0.6	0.7	0.8
10	0.0112	0.0092	0.0065	0.0041	0.0019	0	-0.0020	-0.0037
20	0.0264	0.0213	0.0156	0.0102	0.0049	0	-0.0047	-0.0089
30	0.0450	0.0371	0.0278	0.0184	0.0091	0	-0.0083	-0.0163
40	0.0668	0.0555	0.0423	0.0281	0.0139	0	-0.0130	-0.0255
50	0.0881	0.0737	0.0562	0.0377	0.0188	0	-0.0180	-0.0350
60	0.1058	0.0886	0.0683	0.0454	0.0224	0	-0.0214	-0.0416
	4 wt % ²³⁵U, Actinide & Fission Product Burnup Credit, 10-Year Cooling Time							
	0.1	0.2	0.3	0.4	0.5	0.6	0.7	0.8
10	0.0094	0.0076	0.0055	0.0036	0.0019	0	-0.0015	-0.0028
20	0.0242	0.0198	0.0145	0.0093	0.0044	0	-0.0045	-0.0086
30	0.0432	0.0355	0.0267	0.0176	0.0087	0	-0.0083	-0.0160
40	0.0651	0.0541	0.0410	0.0273	0.0133	0	-0.0127	-0.0251
50	0.0857	0.0716	0.0543	0.0360	0.0178	0	-0.0173	-0.0338
60	0.1019	0.0849	0.0649	0.0431	0.0209	0	-0.0200	-0.0387

Table 15. Sensitivity of Δk_{eff} to reactor water density – 4 wt % (continued)

Burnup (GWd/MTU)	Δk_{eff} Versus Burnup (GWd/MTU) and Depletion Water Density (g/cm^3)							
	4 wt % ^{235}U, Actinide-Only Burnup Credit, 20-Year Cooling Time							
	0.1	0.2	0.3	0.4	0.5	0.6	0.7	0.8
10	0.0109	0.0085	0.0059	0.0039	0.0017	0	-0.0019	-0.0034
20	0.0258	0.0206	0.0152	0.0098	0.0048	0	-0.0045	-0.0084
30	0.0450	0.0366	0.0274	0.0180	0.0089	0	-0.0084	-0.0161
40	0.0670	0.0549	0.0416	0.0275	0.0136	0	-0.0130	-0.0252
50	0.0888	0.0737	0.0559	0.0373	0.0185	0	-0.0178	-0.0347
60	0.1069	0.0887	0.0680	0.0451	0.0226	0	-0.0211	-0.0410
	4 wt % ^{235}U, Actinide & Fission Product Burnup Credit, 20-Year Cooling Time							
	0.1	0.2	0.3	0.4	0.5	0.6	0.7	0.8
10	0.0087	0.0069	0.0048	0.0029	0.0015	0	-0.0017	-0.0028
20	0.0237	0.0191	0.0141	0.0092	0.0045	0	-0.0041	-0.0080
30	0.0431	0.0352	0.0264	0.0174	0.0085	0	-0.0079	-0.0156
40	0.0653	0.0538	0.0406	0.0269	0.0131	0	-0.0127	-0.0245
50	0.0862	0.0713	0.0537	0.0359	0.0177	0	-0.0172	-0.0332
60	0.1023	0.0846	0.0642	0.0425	0.0208	0	-0.0198	-0.0379
	4 wt % ^{235}U, Actinide-Only Burnup Credit, 40-Year Cooling Time							
	0.1	0.2	0.3	0.4	0.5	0.6	0.7	0.8
10	0.0101	0.0081	0.0057	0.0035	0.0017	0	-0.0019	-0.0032
20	0.0253	0.0201	0.0146	0.0095	0.0046	0	-0.0041	-0.0079
30	0.0445	0.0359	0.0268	0.0171	0.0086	0	-0.0081	-0.0156
40	0.0671	0.0550	0.0411	0.0272	0.0136	0	-0.0127	-0.0248
50	0.0897	0.0739	0.0558	0.0372	0.0183	0	-0.0177	-0.0345
60	0.1077	0.0890	0.0679	0.0448	0.0220	0	-0.0209	-0.0407
	4 wt % ^{235}U, Actinide & Fission Product Burnup Credit, 40-Year Cooling Time							
	0.1	0.2	0.3	0.4	0.5	0.6	0.7	0.8
10	0.0083	0.0065	0.0048	0.0032	0.0015	0	-0.0014	-0.0026
20	0.0229	0.0182	0.0133	0.0085	0.0042	0	-0.0043	-0.0077
30	0.0429	0.0348	0.0257	0.0169	0.0082	0	-0.0078	-0.0151
40	0.0654	0.0534	0.0402	0.0265	0.0129	0	-0.0123	-0.0241
50	0.0871	0.0717	0.0539	0.0358	0.0177	0	-0.0168	-0.0323
60	0.1028	0.0842	0.0639	0.0419	0.0201	0	-0.0194	-0.0371

Table 16. Sensitivity of Δk_{eff} to reactor water density – 5 wt %

Burnup (GWd/MTU)	Δk_{eff} Versus Burnup (GWd/MTU) and Depletion Water Density (g/cm^3)							
	5 wt % ^{235}U, Actinide-Only Burnup Credit, 5-Year Cooling Time							
	0.1	0.2	0.3	0.4	0.5	0.6	0.7	0.8
10	0.0075	0.0060	0.0043	0.0027	0.0010	0	-0.0013	-0.0024
20	0.0178	0.0146	0.0108	0.0072	0.0035	0	-0.0031	-0.0059
30	0.0309	0.0254	0.0190	0.0126	0.0060	0	-0.0058	-0.0112
40	0.0470	0.0393	0.0300	0.0199	0.0097	0	-0.0094	-0.0185
50	0.0656	0.0551	0.0426	0.0287	0.0141	0	-0.0140	-0.0274
60	0.0854	0.0724	0.0560	0.0379	0.0188	0	-0.0189	-0.0365
	5 wt % ^{235}U, Actinide & Fission Product Burnup Credit, 5-Year Cooling Time							
	0.1	0.2	0.3	0.4	0.5	0.6	0.7	0.8
10	0.0053	0.0043	0.0032	0.0021	0.0008	0	-0.0009	-0.0019
20	0.0155	0.0125	0.0096	0.0062	0.0030	0	-0.0028	-0.0054
30	0.0294	0.0239	0.0182	0.0119	0.0057	0	-0.0055	-0.0107
40	0.0461	0.0384	0.0292	0.0197	0.0096	0	-0.0092	-0.0181
50	0.0650	0.0546	0.0421	0.0282	0.0139	0	-0.0135	-0.0267
60	0.0841	0.0713	0.0546	0.0369	0.0183	0	-0.0180	-0.0351
	5 wt % ^{235}U, Actinide-Only Burnup Credit, 10-Year Cooling Time							
	0.1	0.2	0.3	0.4	0.5	0.6	0.7	0.8
10	0.0074	0.0059	0.0042	0.0029	0.0012	0	-0.0012	-0.0023
20	0.0172	0.0138	0.0101	0.0066	0.0030	0	-0.0031	-0.0057
30	0.0305	0.0251	0.0189	0.0125	0.0059	0	-0.0054	-0.0106
40	0.0467	0.0387	0.0293	0.0195	0.0095	0	-0.0093	-0.0181
50	0.0661	0.0551	0.0425	0.0286	0.0143	0	-0.0137	-0.0270
60	0.0862	0.0725	0.0557	0.0378	0.0186	0	-0.0185	-0.0363
	5 wt % ^{235}U, Actinide & Fission Product Burnup Credit, 10-Year Cooling Time							
	0.1	0.2	0.3	0.4	0.5	0.6	0.7	0.8
10	0.0049	0.0041	0.0030	0.0018	0.0010	0	-0.0008	-0.0018
20	0.0151	0.0123	0.0091	0.0059	0.0030	0	-0.0025	-0.0052
30	0.0287	0.0234	0.0178	0.0117	0.0057	0	-0.0054	-0.0105
40	0.0461	0.0384	0.0292	0.0195	0.0097	0	-0.0091	-0.0175
50	0.0654	0.0547	0.0418	0.0281	0.0139	0	-0.0136	-0.0266
60	0.0848	0.0714	0.0547	0.0367	0.0183	0	-0.0176	-0.0349

Table 16. Sensitivity of Δk_{eff} to reactor water density – 5 wt % (continued)

Burnup (GWd/MTU)	Δk_{eff} Versus Burnup (GWd/MTU) and Depletion Water Density (g/cm^3)							
	5 wt % ^{235}U, Actinide-Only Burnup Credit, 20-Year Cooling Time							
	0.1	0.2	0.3	0.4	0.5	0.6	0.7	0.8
10	0.0072	0.0056	0.0039	0.0023	0.0010	0	-0.0010	-0.0022
20	0.0169	0.0133	0.0097	0.0063	0.0030	0	-0.0028	-0.0052
30	0.0301	0.0243	0.0180	0.0119	0.0057	0	-0.0053	-0.0104
40	0.0472	0.0388	0.0295	0.0194	0.0096	0	-0.0087	-0.0171
50	0.0665	0.0550	0.0421	0.0283	0.0141	0	-0.0135	-0.0263
60	0.0869	0.0730	0.0559	0.0377	0.0188	0	-0.0181	-0.0358
	5 wt % ^{235}U, Actinide & Fission Product Burnup Credit, 20-Year Cooling Time							
	0.1	0.2	0.3	0.4	0.5	0.6	0.7	0.8
10	0.0047	0.0037	0.0027	0.0020	0.0008	0	-0.0008	-0.0017
20	0.0145	0.0118	0.0087	0.0058	0.0026	0	-0.0024	-0.0048
30	0.0283	0.0229	0.0171	0.0111	0.0054	0	-0.0052	-0.0102
40	0.0460	0.0379	0.0287	0.0191	0.0094	0	-0.0087	-0.0171
50	0.0659	0.0548	0.0415	0.0280	0.0136	0	-0.0134	-0.0259
60	0.0856	0.0712	0.0542	0.0364	0.0181	0	-0.0176	-0.0343
	5 wt % ^{235}U, Actinide-Only Burnup Credit, 40-Year Cooling Time							
	0.1	0.2	0.3	0.4	0.5	0.6	0.7	0.8
10	0.0068	0.0052	0.0038	0.0024	0.0012	0	-0.0008	-0.0017
20	0.0162	0.0127	0.0093	0.0060	0.0029	0	-0.0023	-0.0048
30	0.0298	0.0239	0.0173	0.0113	0.0055	0	-0.0053	-0.0098
40	0.0469	0.0381	0.0283	0.0190	0.0090	0	-0.0089	-0.0170
50	0.0672	0.0552	0.0420	0.0280	0.0138	0	-0.0131	-0.0258
60	0.0878	0.0731	0.0556	0.0374	0.0184	0	-0.0180	-0.0353
	5 wt % ^{235}U, Actinide & Fission Product Burnup Credit, 40-Year Cooling Time							
	0.1	0.2	0.3	0.4	0.5	0.6	0.7	0.8
10	0.0044	0.0036	0.0025	0.0017	0.0009	0	-0.0006	-0.0012
20	0.0139	0.0110	0.0080	0.0052	0.0022	0	-0.0022	-0.0046
30	0.0280	0.0227	0.0168	0.0111	0.0053	0	-0.0047	-0.0091
40	0.0460	0.0374	0.0282	0.0185	0.0089	0	-0.0086	-0.0167
50	0.0659	0.0543	0.0411	0.0273	0.0132	0	-0.0133	-0.0255
60	0.0863	0.0715	0.0543	0.0363	0.0179	0	-0.0171	-0.0335

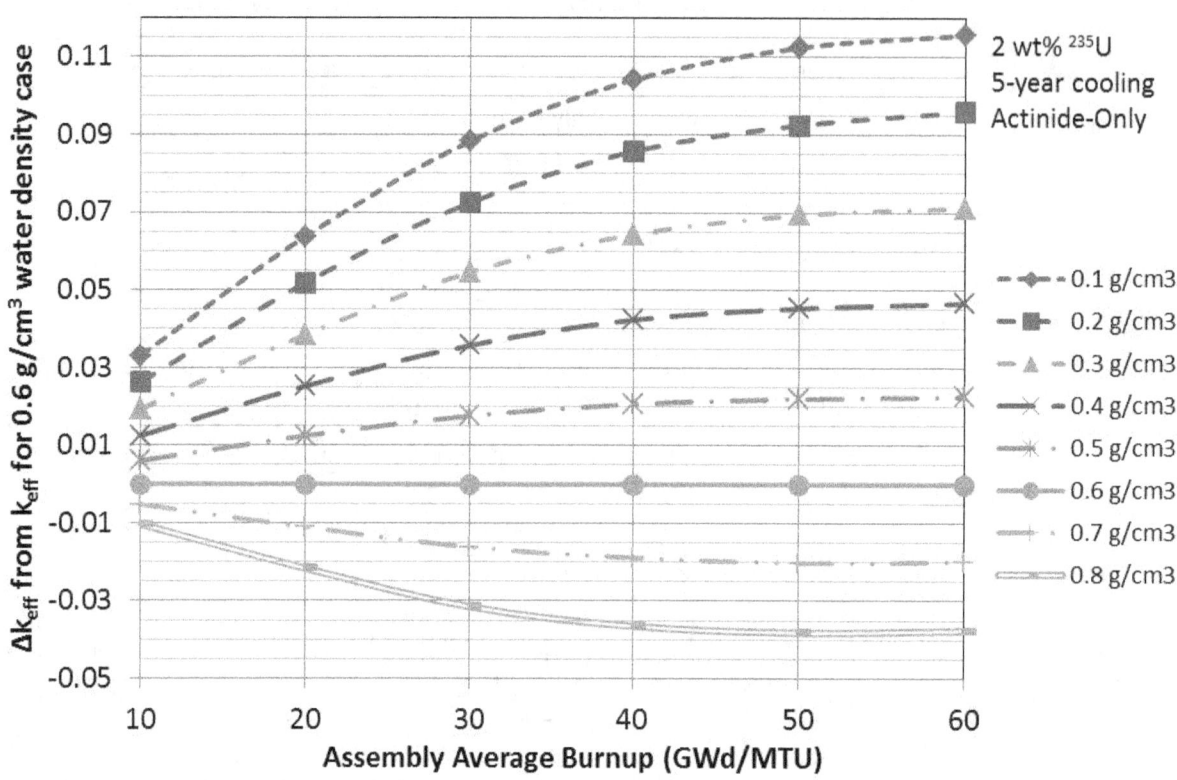

Figure 21. Sensitivity of △k to ρ_mod (2 wt %, 5-year cooling time, actinide-only).

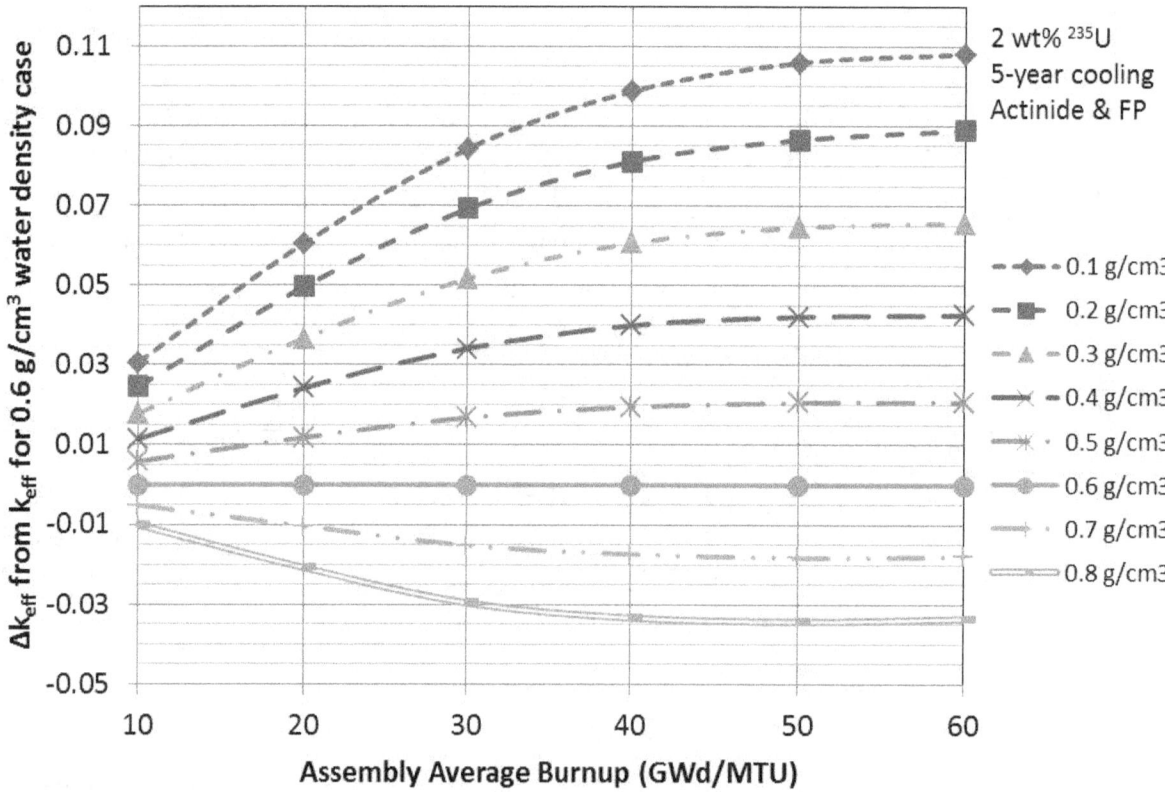

Figure 22. Sensitivity of △k to ρ_mod (2 wt %, 5-year cooling time, actinide+16FP).

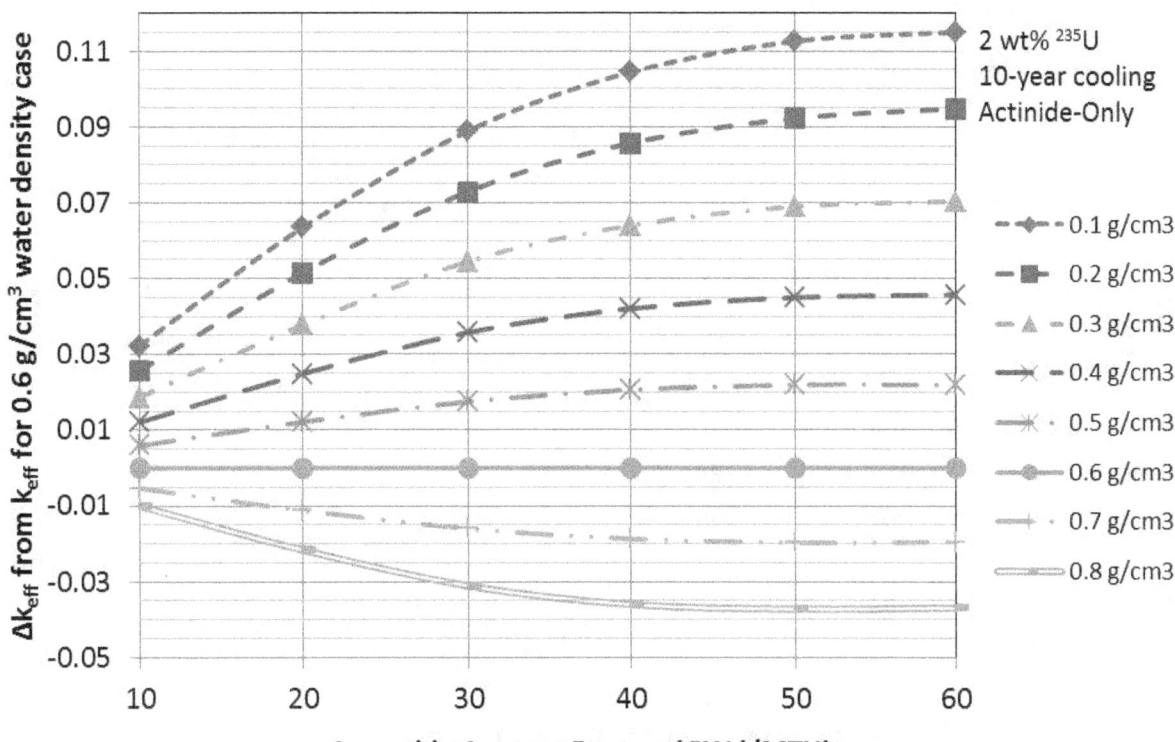

Figure 23. Sensitivity of △k to ρ$_{mod}$ (2 wt %, 10-year cooling time, actinide-only).

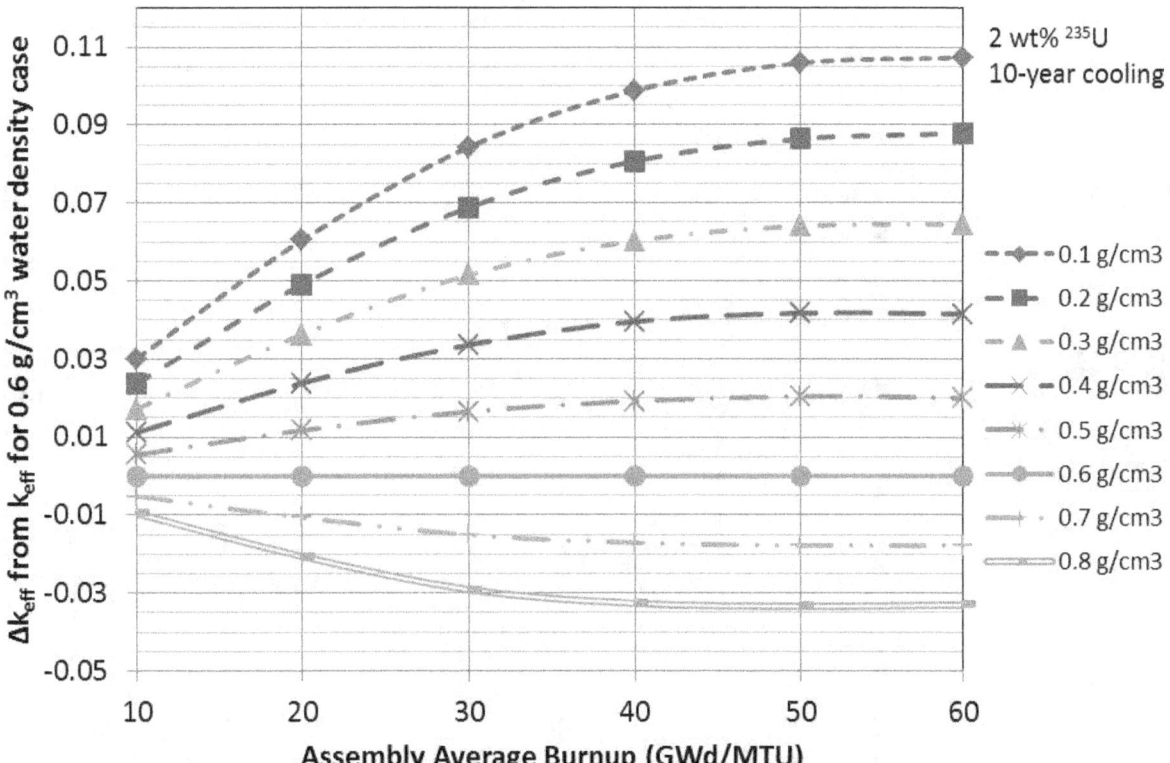

Figure 24. Sensitivity of △k to ρ$_{mod}$ (2 wt %, 10-year cooling time, actinide+16FP).

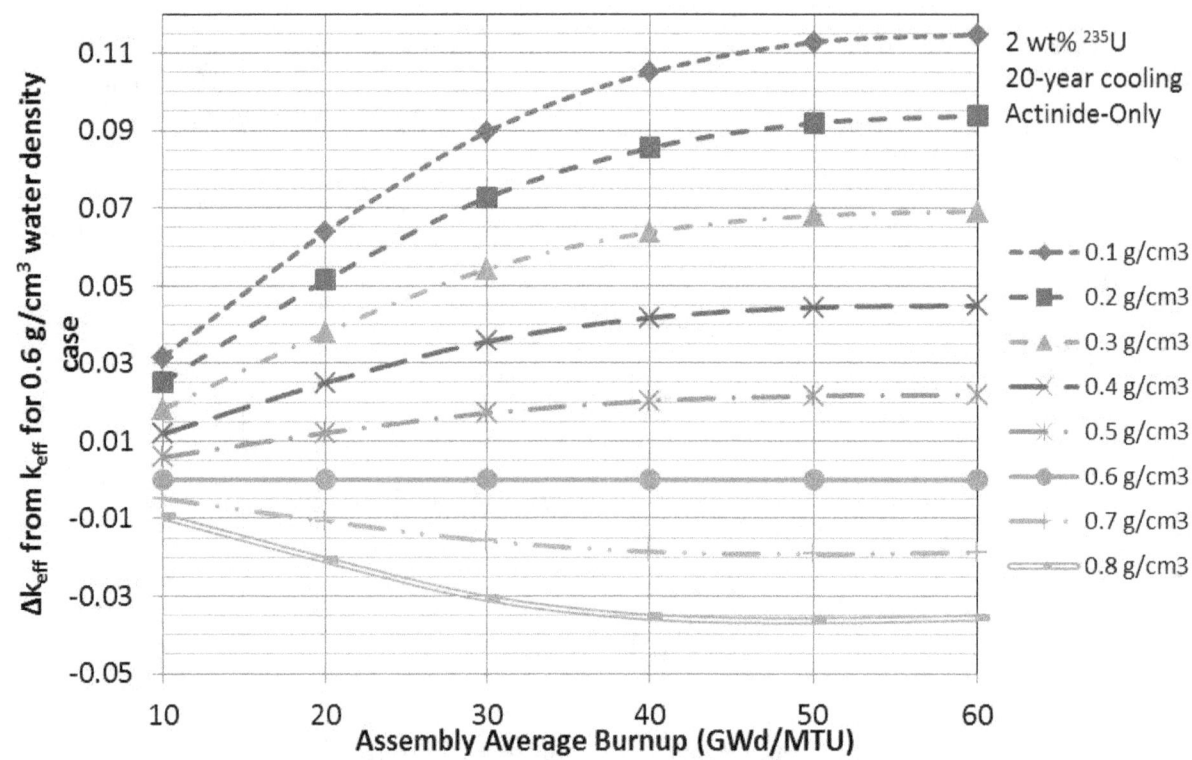

Figure 25. Sensitivity of △k to ρ_mod (2 wt %, 20-year cooling time, actinide-only).

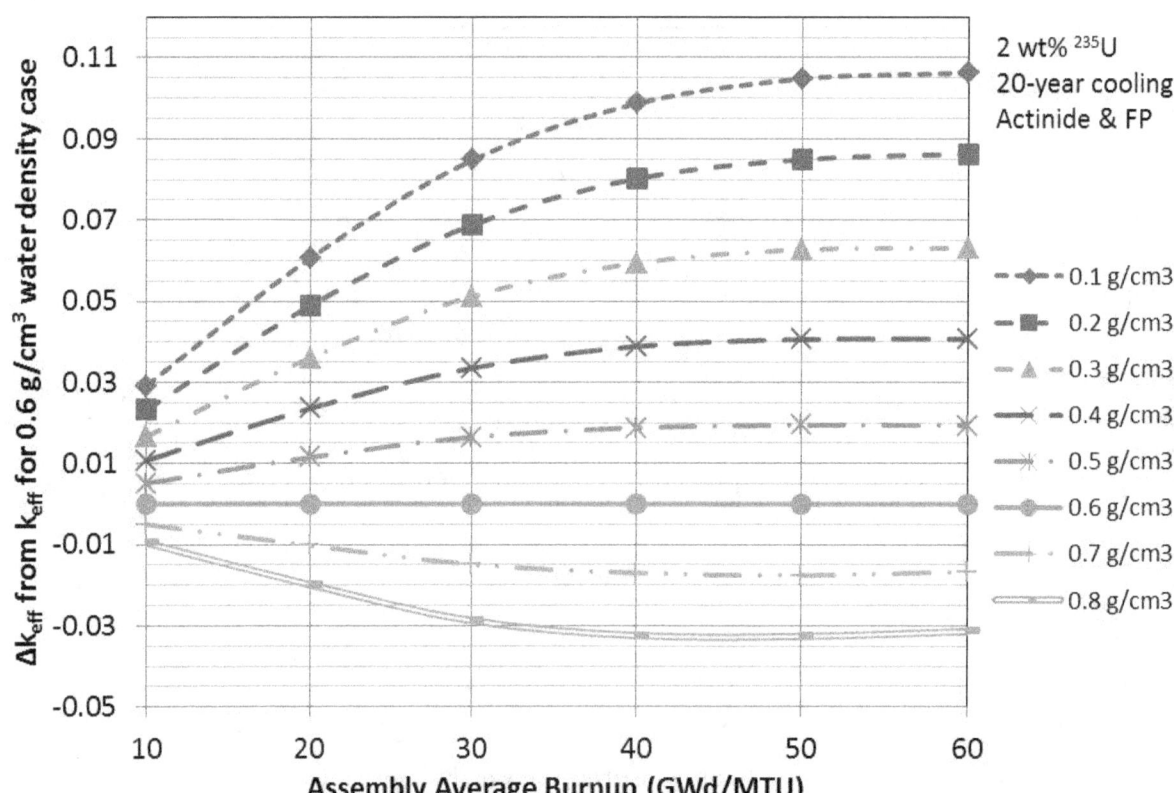

Figure 26. Sensitivity of △k to ρ_mod (2 wt %, 20-year cooling time, actinide+16FP).

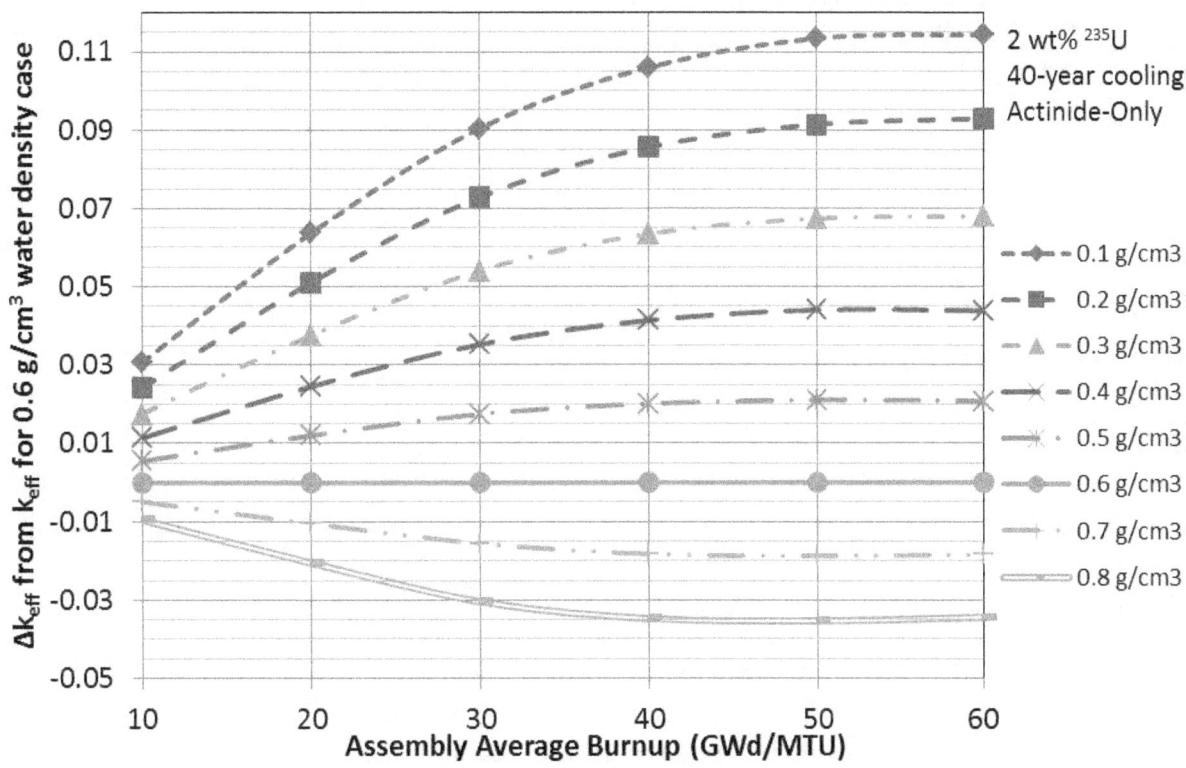

Figure 27. Sensitivity of Δk to ρ$_{mod}$ (2 wt %, 40-year cooling time, actinide-only).

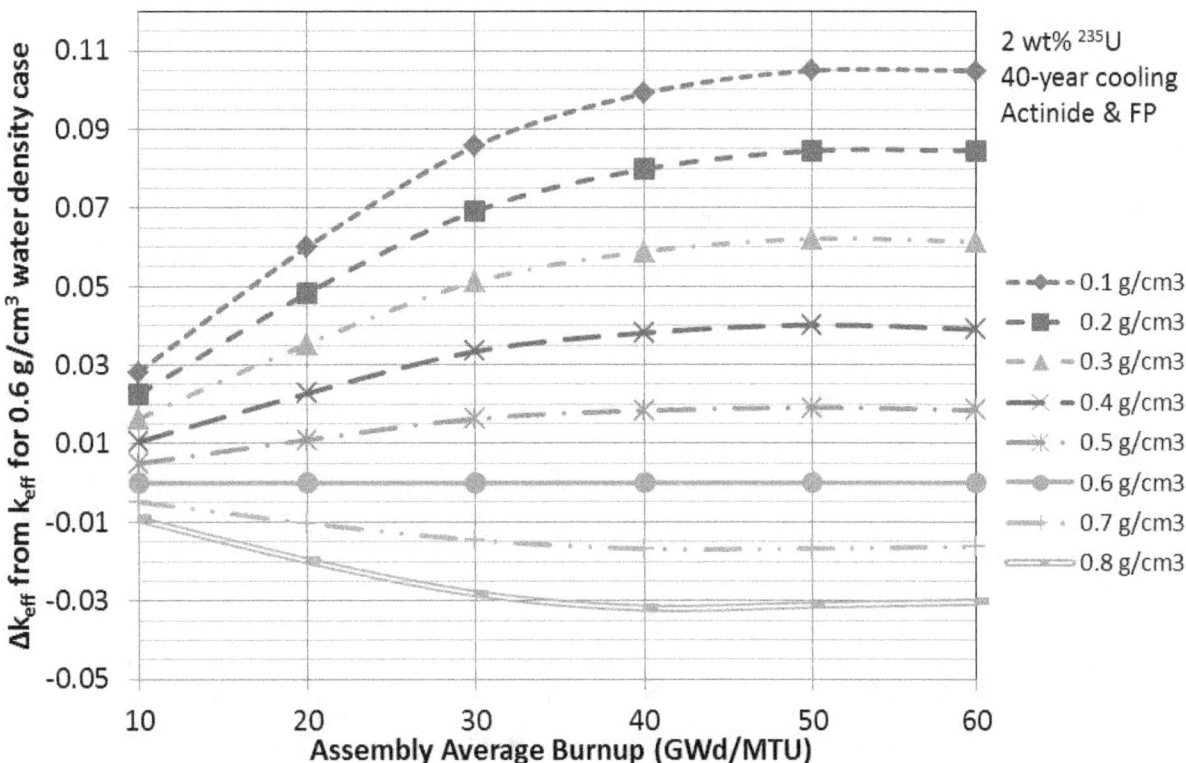

Figure 28. Sensitivity of Δk to ρ$_{mod}$ (2 wt %, 40-year cooling time, actinide+16FP).

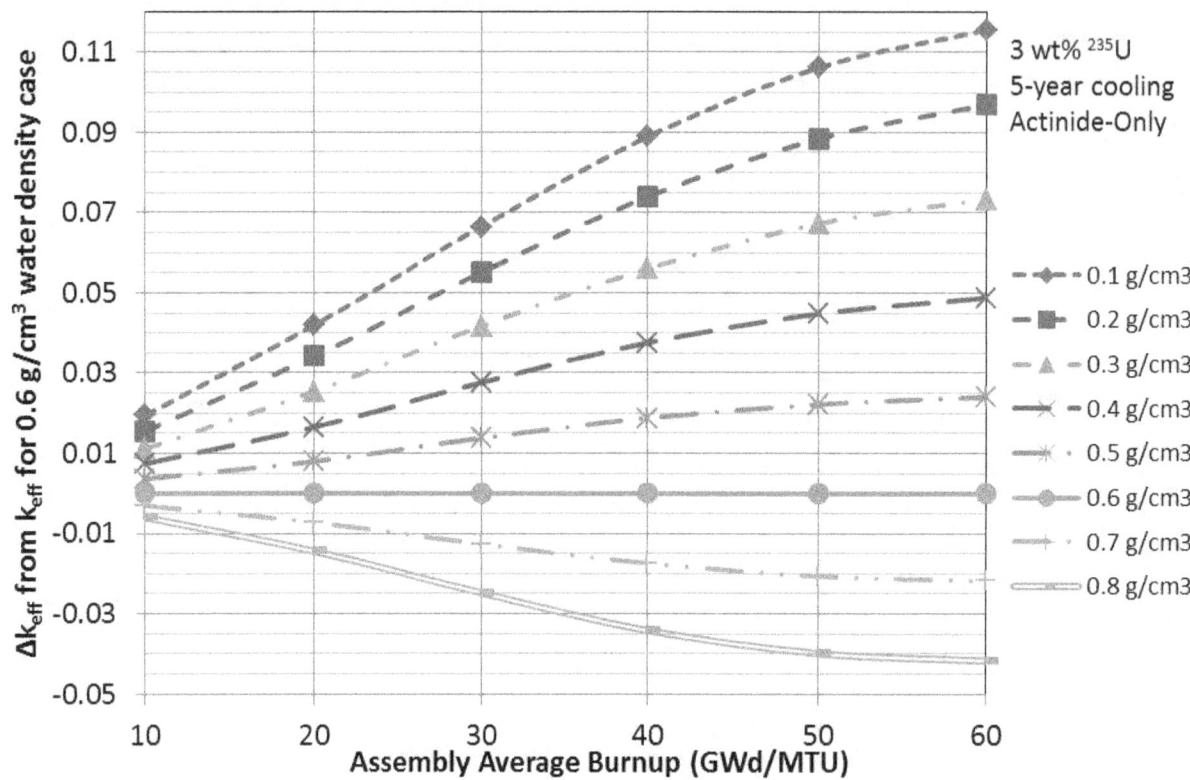

Figure 29. Sensitivity of Δk to ρ_{mod} (3 wt %, 5-year cooling time, actinide-only).

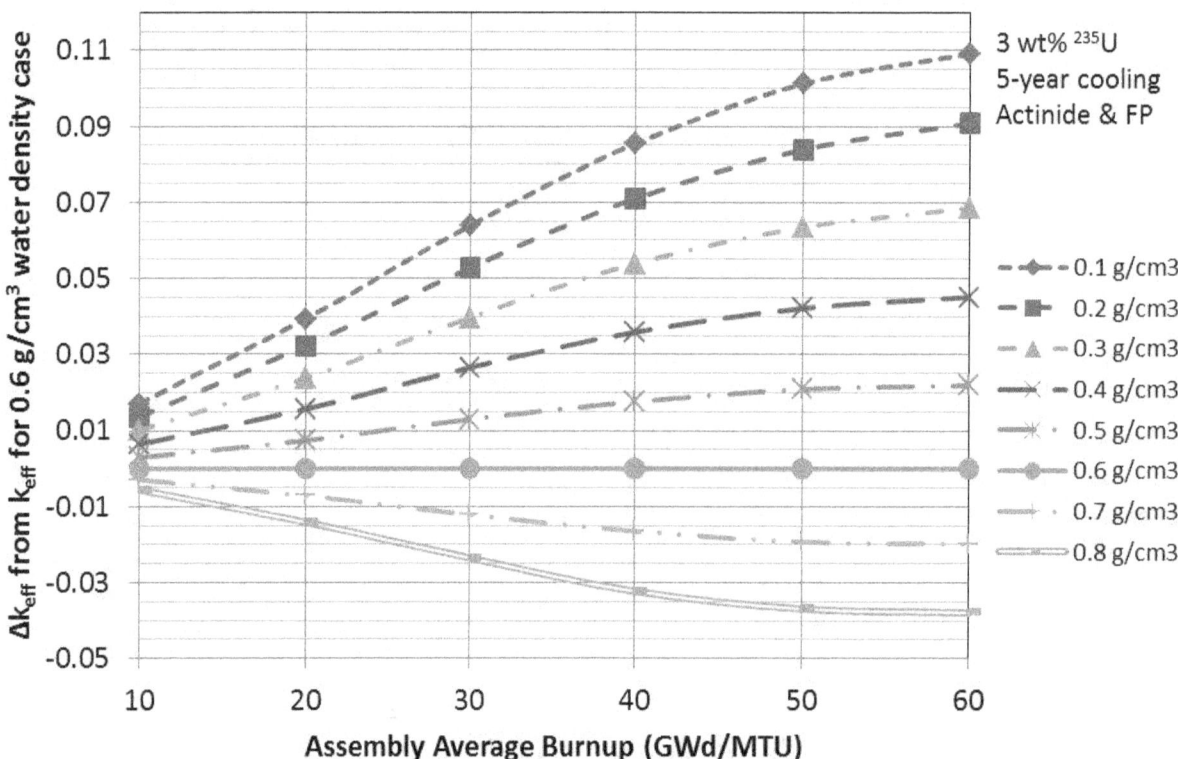

Figure 30. Sensitivity of Δk to ρ_{mod} (3 wt %, 5-year cooling time, actinide+16FP).

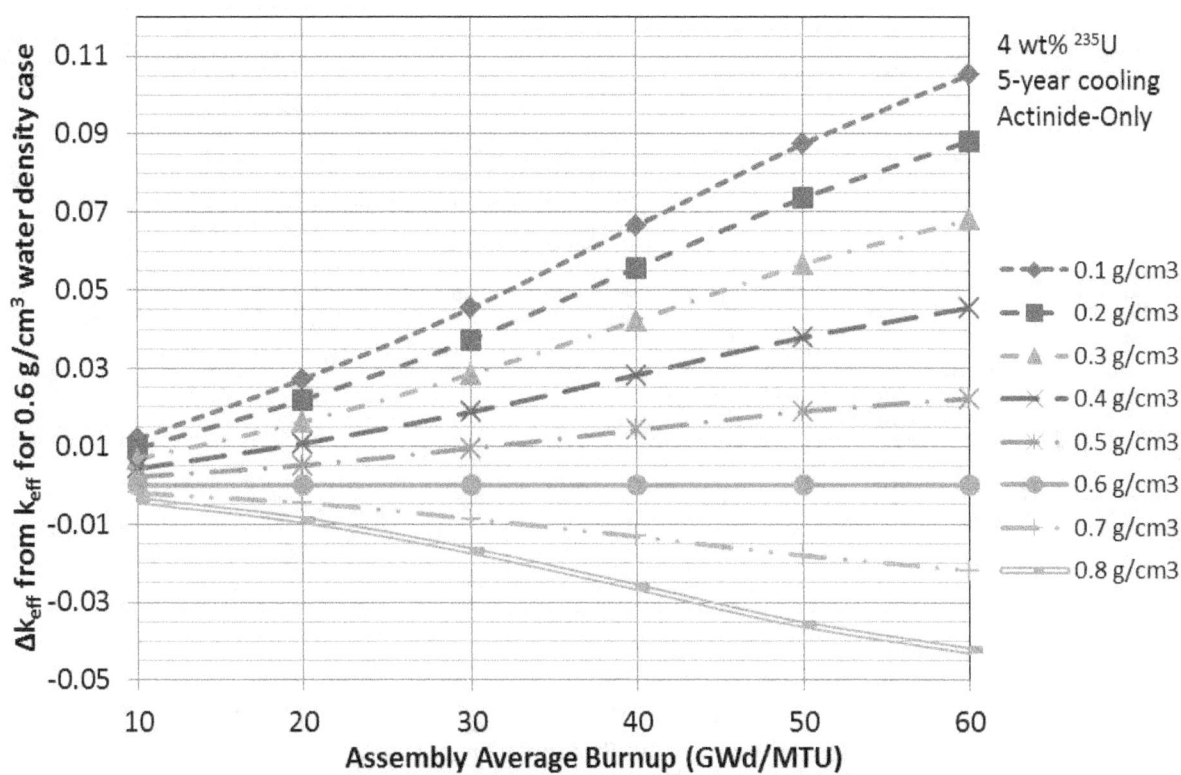

Figure 31. Sensitivity of Δk to ρ_{mod} (4 wt %, 5-year cooling time, actinide-only).

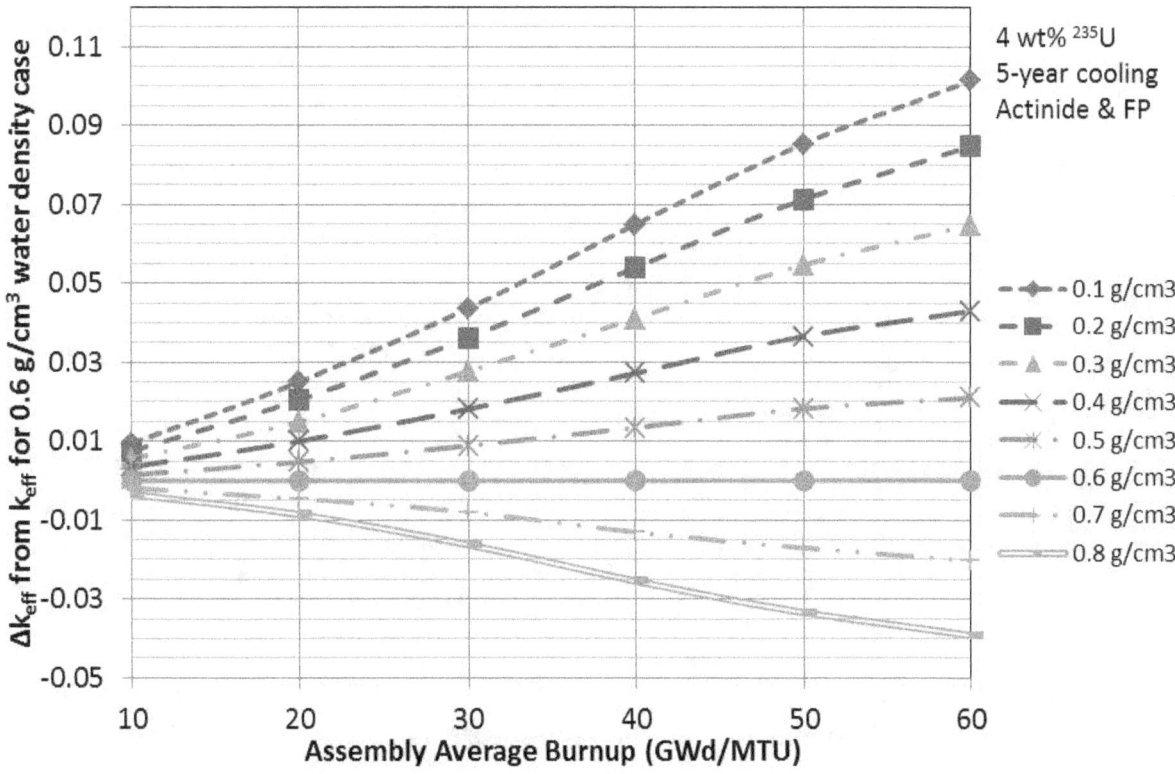

Figure 32. Sensitivity of Δk to ρ_{mod} (4 wt %, 5-year cooling time, actinide+16FP).

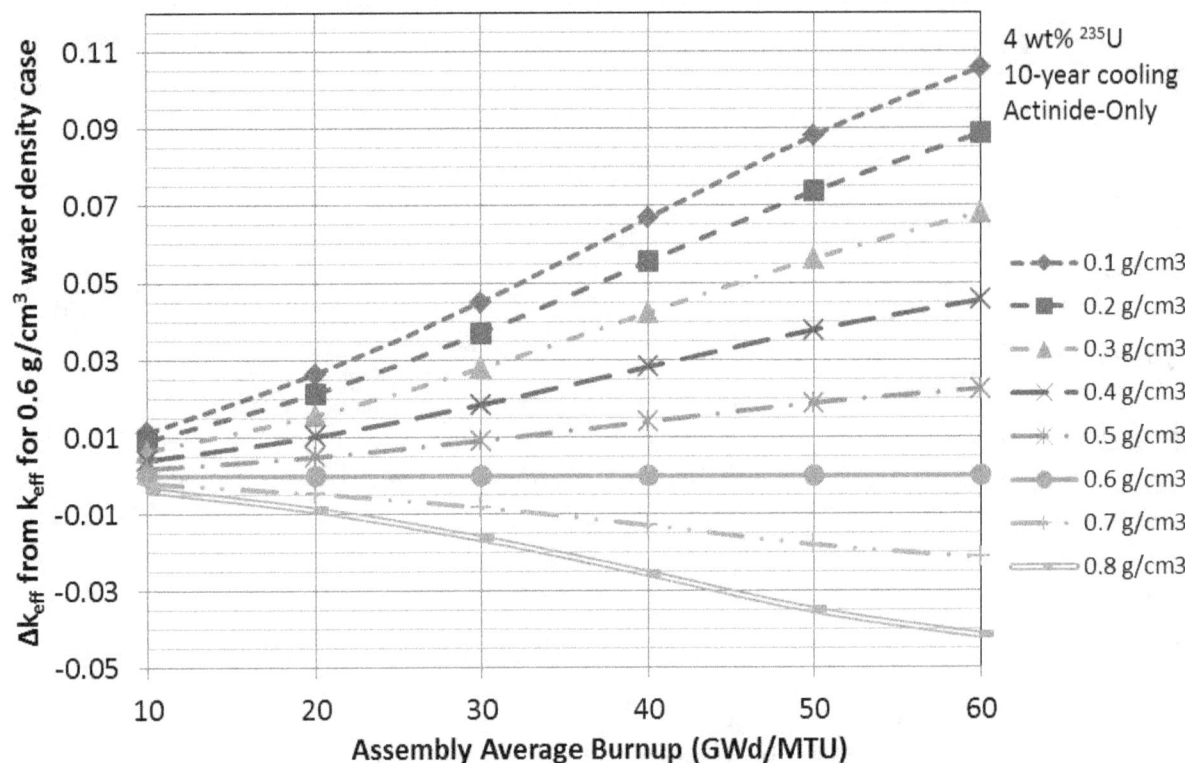

Figure 33. Sensitivity of Δk to ρ_{mod} (4 wt %, 10-year cooling time, actinide-only).

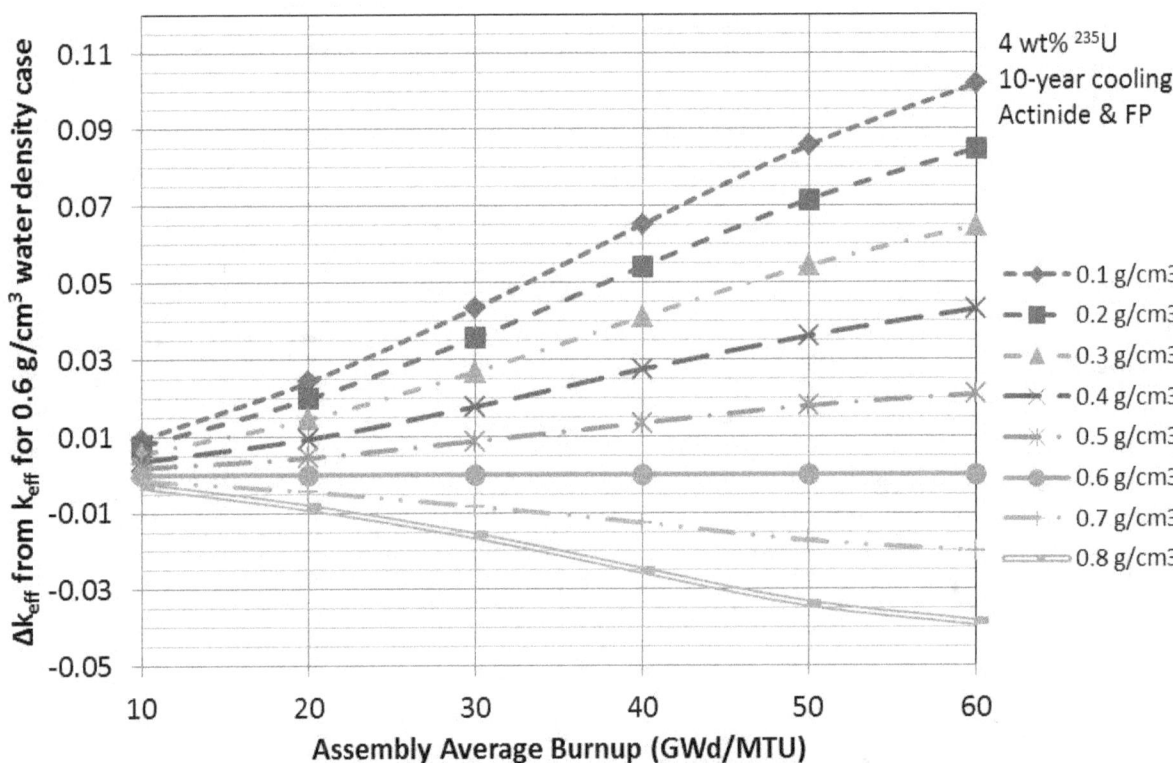

Figure 34. Sensitivity of Δk to ρ_{mod} (4 wt %, 10-year cooling time, actinide+16FP).

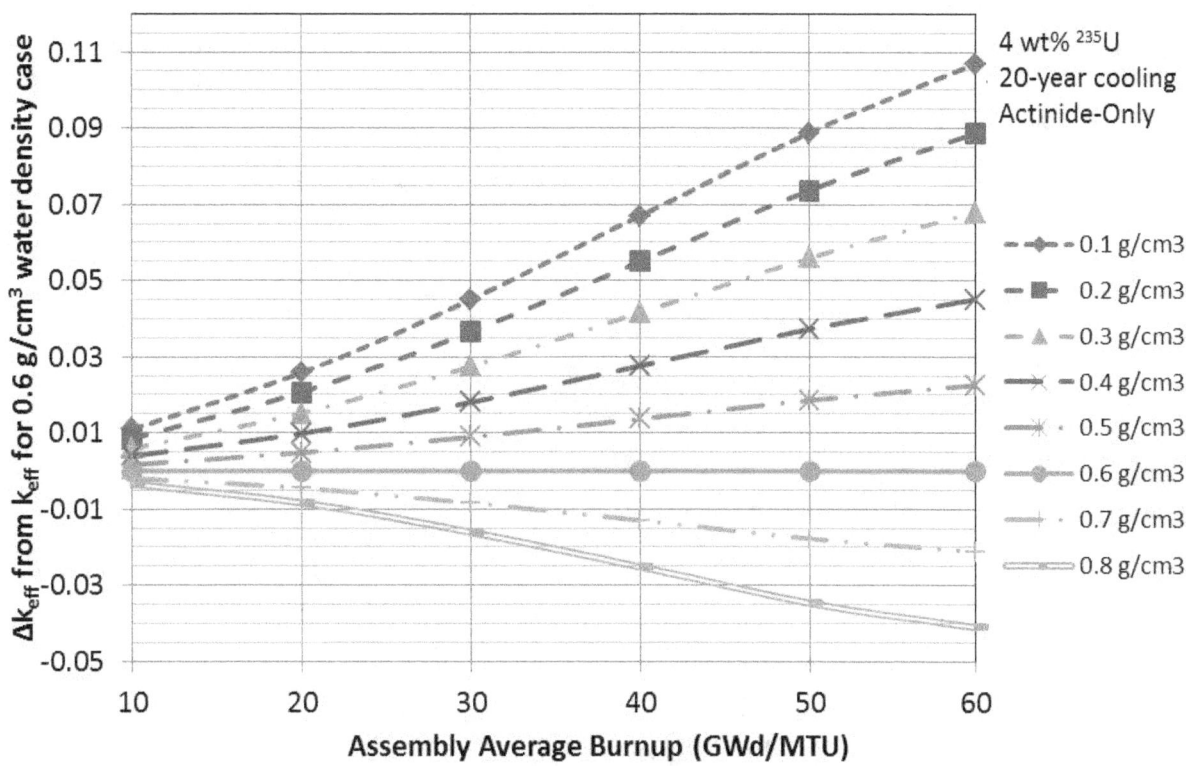

Figure 35. Sensitivity of Δk to ρ_{mod} (4 wt %, 20-year cooling time, actinide-only).

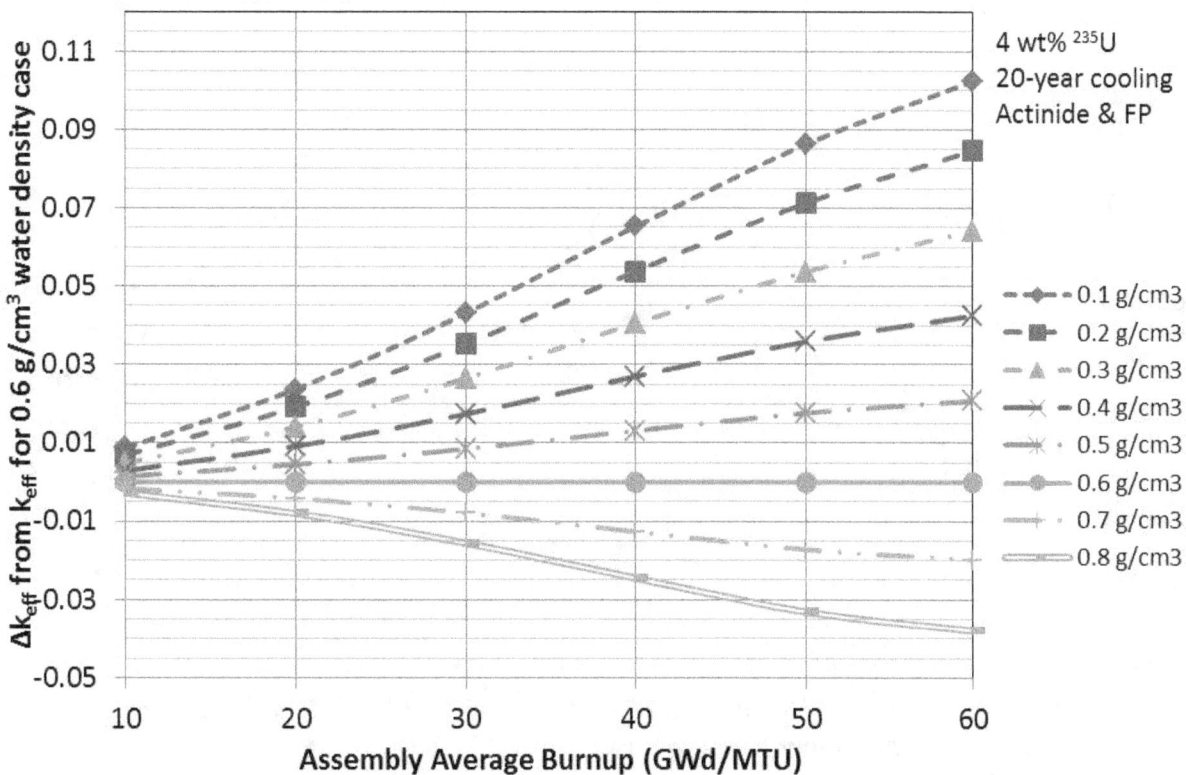

Figure 36. Sensitivity of Δk to ρ_{mod} (4 wt %, 20-year cooling time, actinide+16FP).

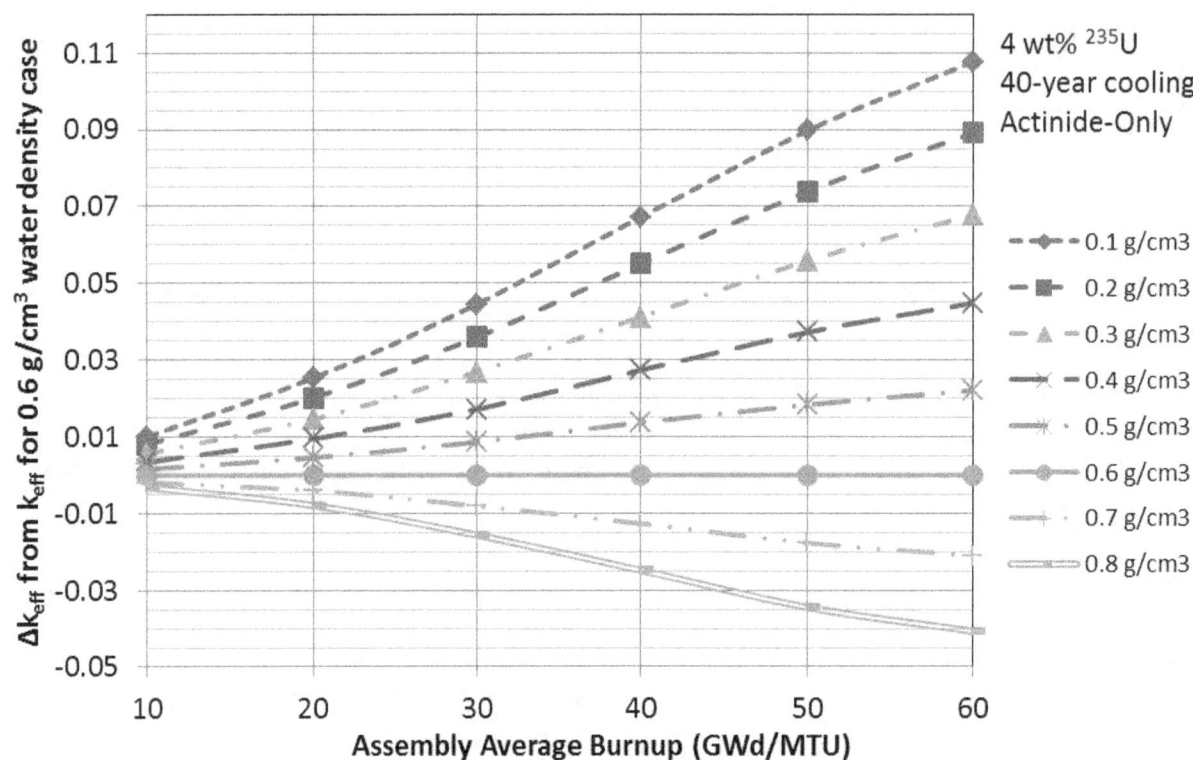

Figure 37. Sensitivity of △k to ρ_mod (4 wt %, 40-year cooling time, actinide-only).

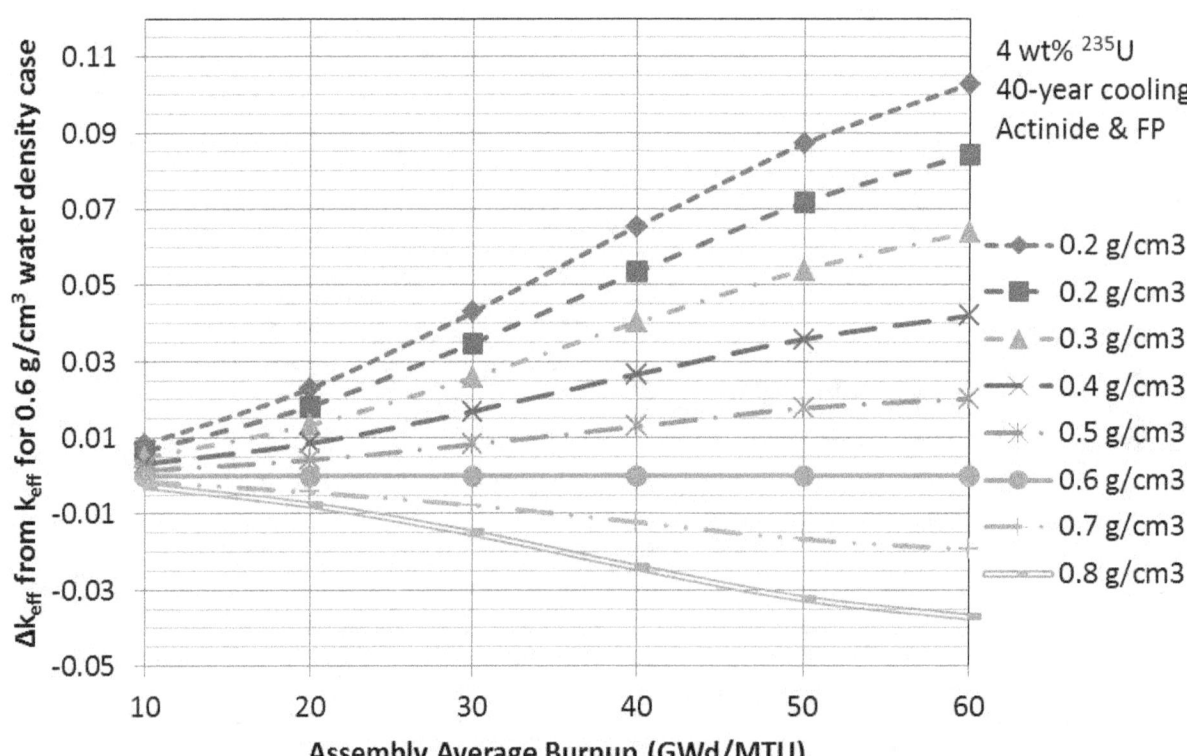

Figure 38. Sensitivity of △k to ρ_mod (4 wt %, 40-year cooling time, actinide+16FP).

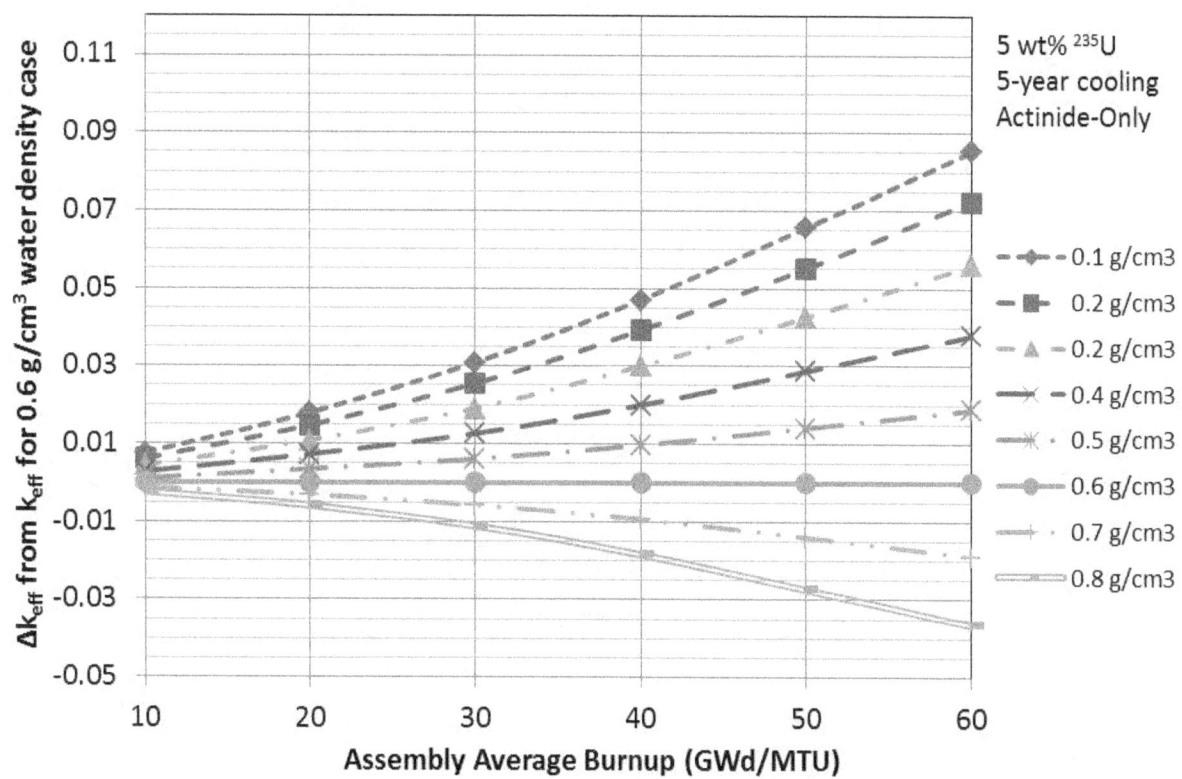

Figure 39. Sensitivity of △k to ρ_mod (5 wt %, 5-year cooling time, actinide-only).

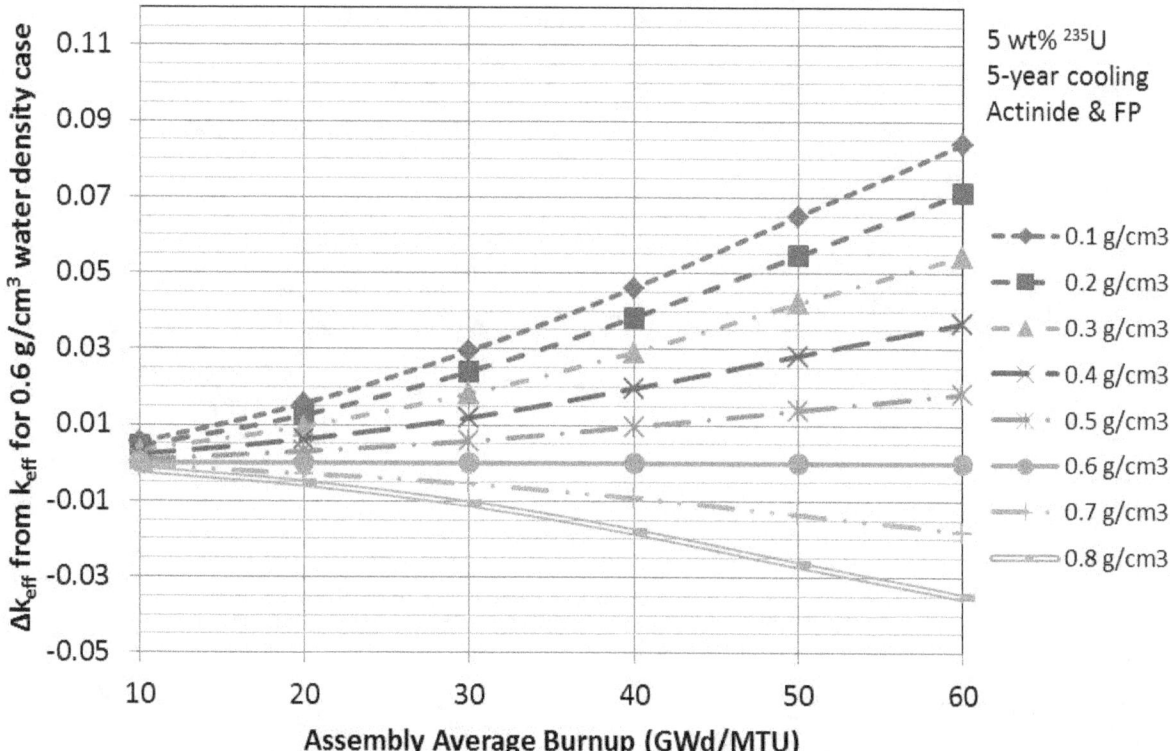

Figure 40. Sensitivity of △k to ρ_mod (5 wt %, 5-year cooling time, actinide+16FP).

NRC FORM 335 (12-2010) NRCMD 3.7	U.S. NUCLEAR REGULATORY COMMISSION	1. REPORT NUMBER (Assigned by NRC, Add Vol., Supp., Rev., and Addendum Numbers, if any.)
BIBLIOGRAPHIC DATA SHEET (See instructions on the reverse)		NUREG/CR-7158 (ORNL/TM-2012/261)

2. TITLE AND SUBTITLE		3. DATE REPORT PUBLISHED	
Review and Prioritization of Technical Issues Related to Burnup Credit for BWR Fuel		MONTH	YEAR
		02	2013
		4. FIN OR GRANT NUMBER	
		V6061	

5. AUTHOR(S)	6. TYPE OF REPORT
D. E. Mueller S. M. Bowman W. J. Marshall J. M. Scaglione	Technical
	7. PERIOD COVERED *(Inclusive Dates)*

8. PERFORMING ORGANIZATION - NAME AND ADDRESS *(If NRC, provide Division, Office or Region, U.S. Nuclear Regulatory Commission, and mailing address; if contractor, provide name and mailing address.)*

Oak Ridge National Laboratory
P.O Box 2008, MS-6170
Oak Ridge, TN 37831-6170

9. SPONSORING ORGANIZATION - NAME AND ADDRESS *(If NRC, type "Same as above"; if contractor, provide NRC Division, Office or Region, U.S. Nuclear Regulatory Commission, and mailing address.)*

Division of Systems Analysis, Office of Nuclear Regulatory Research
U.S. Nuclear Regulatory Commission
Washington, DC 20555-0001

10. SUPPLEMENTARY NOTES
Mourad Aissa, NRC Project Manager

11. ABSTRACT *(200 words or less)*

The work presented in this report is primarily a sensitivity study designed to identify and rank phenomena and parameters important to BWR burnup credit methodology. This work is an extension of the work reported in NUREG/CR-XXXX, Computational Benchmark for Estimated Reactivity Margin from Fission Products and Minor Actinides in BWR Burnup Credit, which defines the baseline BWR spent fuel cask model used in this report and provides estimates for the reactivity margin associated with fission products and minor actinides. All calculations supporting this work were performed using the SCALE 6.1 code system with the 238-neutron energy group ENDF/B-VII-based nuclear data library.

Discussion of and recommendations for future work supporting implementation of BWR burnup credit beyond the currently used peak reactivity method are provided. A high priority is recommended for development of guidance for identification and use of axial burnup distribution data, for treatment of axial moderator density distributions, and for treatment of control blade usage during depletion calculations. In addition to these higher priority items, several medium and lower priority items are identified and discussed

12. KEY WORDS/DESCRIPTORS *(List words or phrases that will assist researchers in locating the report.)*	13. AVAILABILITY STATEMENT
burnup credit BWR fission products actinides sensitivity	unlimited
	14. SECURITY CLASSIFICATION
	(This Page) unclassified
	(This Report) unclassified
	15. NUMBER OF PAGES
	16. PRICE

NRC FORM 335 (12-2010)

UNITED STATES
NUCLEAR REGULATORY COMMISSION
WASHINGTON, DC 20555-0001

OFFICIAL BUSINESS

NUREG/CR-7158

Review and Prioritization of Technical Issues Related to Burnup Credit for BWR Fuel

February 2013